Springerbriefs in Research and Innovation Governance

Editor-in-Chief

Doris Schroeder, Centre for Professional Ethics, University of Central Lancashire, Preston, Lancashire, UK

Konstantinos Iatridis, School of Management, University of Bath, Bath, UK

SpringerBriefs in Research and Innovation Governance present concise summaries of cutting-edge research and practical applications across a wide spectrum of governance activities that are shaped and informed by, and in turn impact research and innovation, with fast turnaround time to publication. Featuring compact volumes of 50 to 125 pages, the series covers a range of content from professional to academic. Monographs of new material are considered for the SpringerBriefs in Research and Innovation Governance series. Typical topics might include: a timely report of state-of-the-art analytical techniques, a bridge between new research results, as published in journal articles and a contextual literature review, a snapshot of a hot or emerging topic, an in-depth case study or technical example, a presentation of core concepts that students and practitioners must understand in order to make independent contributions, best practices or protocols to be followed, a series of short case studies/debates highlighting a specific angle. SpringerBriefs in Research and Innovation Governance allow authors to present their ideas and readers to absorb them with minimal time investment. Both solicited and unsolicited manuscripts are considered for publication.

More information about this series at http://www.springer.com/series/13811

Doris Schroeder • Kate Chatfield
Michelle Singh • Roger Chennells
Peter Herissone-Kelly

Equitable Research Partnerships

A Global Code of Conduct to Counter Ethics Dumping

Foreword by Klaus Leisinger

OPEN

 Springer

Doris Schroeder
Centre for Professional Ethics
University of Central Lancashire
Preston, Lancashire, UK

Kate Chatfield
Centre for Professional Ethics
University of Central Lancashire
Preston, Lancashire, UK

Michelle Singh
Africa Office
European & Developing Countries Clinical
Trials Partnership
Cape Town, South Africa

Roger Chennells
Chennells Albertyn Attorneys
Stellenbosch, South Africa

Peter Herissone-Kelly
School of Humanities and Social Sciences
University of Central Lancashire
Preston, Lancashire, UK

ISSN 2452-0519   ISSN 2452-0527   (electronic)
SpringerBriefs in Research and Innovation Governance
ISBN 978-3-030-15744-9   ISBN 978-3-030-15745-6   (eBook)
https://doi.org/10.1007/978-3-030-15745-6

This Springer imprint is published by the registered company Springer Nature Switzerland AG.
The registered company address is: Gewerbestrasse 11, 6330 Cham, Switzerland

To Reverend Mario Mahongo (1952–2018)

Foreword

In September 2015, after intensive public consultation, the international community went on record with a plan of action for people, planet and prosperity: the *2030 Agenda for Sustainable Development*. All country representatives and all stakeholders expressed their determination

> to take the bold and transformative steps which are urgently needed to shift the world on to a sustainable and resilient path. As we embark on this collective journey, *we pledge that no one will be left behind* [emphasis added]. (UN 2015).

To stimulate action, the heads of states and governments defined 17 sustainable development goals and 169 targets to be achieved by 2030. Successes in efforts to end extreme poverty, achieve food security and ensure healthy lives, as well as successes towards all other goals, depend not only on goal-oriented societal reforms and the mobilization of substantial financial and technical assistance, but also on significant technological, biomedical and other innovations.

Ensuring the success of the Agenda 2030 requires massive research and development efforts as well new forms of research co-creation on a level playing field and with a universal professional ethos.

Leaving no one behind does not "only" include reducing income and wealth inequalities, and affirmative action in support of better opportunities for self-determined living within and among countries. It also implies reaching those most at risk from poverty and its impacts. This again necessitates research focused on the needs of the poor in a way that does not infringe their human rights.

Research and innovation can only be sustainably successful when based on societal trust. The precondition for societal trust and public acceptance is the perception that work is done with integrity and based on fundamental values shared by the global community. Trust depends not only on research work being compliant with laws and regulations, but also, more than ever, on its legitimacy.

Such legitimacy can be achieved through inclusion and, importantly, the co-design of solutions with vulnerable populations. *Leaving no one behind* also means leaving no one behind *throughout* the research process, aiming for research *with*, not *about*, vulnerable populations.

The results of the TRUST Project, whose Global Code of Conduct for Research in Resource-Poor Settings (GCC) this book celebrates, contribute to realizing the European Union's ambition of a more inclusive, equal and sustainable global society – a profound expectation of people all over the world.

The fact that the GCC now exists and has been welcomed by the European Commission as a precondition for its research grants is only a beginning.

My hope is that enlightened stakeholders in public institutions, foundations and the private sector will now start a discourse and apply moral imagination to the concrete consequences of the GCC. This relates to the processes and content of their research endeavours as well as the selection criteria for hiring, promoting and remunerating the research workforce.

Research excellence is no longer only defined by playing by the rules and being "successful". The results of discourses about the operationalization of the TRUST values of *fairness, respect, care* and *honesty* are the new benchmark for excellence.

Basel, Switzerland   Klaus Leisinger

Reference

UN (2015) Transforming our world: the 2030 agenda for sustainable development. United Nations. https://www.un.org/development/desa/dspd/2015/08/transforming-our-world-the-2030-agenda-for-sustainable-development/

Professor Klaus Leisinger, a social scientist and economist, is the President of the Global Values Alliance in Basel, Switzerland. He served as an adviser on corporate responsibility to UN Secretaries-General Kofi Annan and Ban Ki-moon. He is currently a member of the Leadership Council of the UN Sustainable Development Solutions Network. In 2011, he was awarded the first ever Outstanding Contribution to Global Health Award by South-South Awards for his successful work on eradicating leprosy.

Acknowledgements

Writing a book is child's play compared to writing a new ethics code – a monumental task achieved by the 56 individuals named in the Appendix as the proud authors of the Global Code of Conduct for Research in Resource-Poor Settings (GCC). Thus, by the time we started writing this book, the bulk of the work had already been done.

The task of conveying the collective pride of these 56 authors to the world was entrusted to the Reverend Mario Mahongo, an honoured San Leader born in Angola. He was due to travel from the Kalahari Desert to Stockholm, Sweden, in May 2018 to launch the GCC. Just one day before flying to Europe, he died in a car crash. This book is dedicated to Mario. His last recorded statement about research ethics was: "I don't want researchers to see us as museums who cannot speak for themselves and who don't expect something in return" (Chapter 7). This statement expresses the fairness element of the GCC beautifully.

The GCC was produced by the TRUST project, an initiative funded by the European Commission (EC) Horizon 2020 Programme, agreement number 664771.

Dorian Karatzas, Roberta Monachello, Dr Louiza Kalokairinou, Edyta Sikorska, Yves Dumont and Wolfgang Bode formed the magnificent EC team supporting the TRUST project.

Thanks to Dorian for ensuring that the GCC was brought to the attention of the highest level of decision-making on ethics in the EC, for suggesting TRUST as a research and development success story of Horizon 2020 (EC 2018) and for having the GCC checked by the EC legal department in time for our event at the European Parliament in June 2018. Without Dorian's efforts, the code would not have the standing it has now, as a mandatory reference document for EC framework programmes.

Thanks to Roberta for believing in our work and for being a most enthusiastic, supportive and interested project officer, despite several amendments. Thanks to Wolfgang for facilitating one of those amendments very professionally and in record time during a summer break.

Thanks to Louiza for providing insightful funder input during the GCC development phase. Thanks to Edyta for organizing a very stimulating training event for EC staff on the GCC. Thanks to Yves Dumont for inventing the term "ethics dumping" in 2013.

Thanks to Stelios Kouloglou, MEP, and Dr Mihalis Kritikos for giving us the opportunity to present the GCC at a European Parliament event.

Thanks to Dr Wolfgang Burtscher, the EC's deputy director-general for Research and Innovation, for announcing in person at the European Parliament event that the GCC would henceforth be a mandatory reference document for EC framework programmes.

Thanks to the University of Cape Town for being the first university to adopt the GCC in April 2019. This is owed to Prof. Rachel Wynberg's long-term commitment to equitable research partnerships and the protection of vulnerable populations in research.

Thanks to Joyce Adhiambo Odhiambo and her colleagues in Nairobi for preparing the excellent speech on the four values of the GCC – fairness, respect, care and honesty – that she presented at the European Parliament (TRUST 2018).

Thanks to Leana Snyders, the director of the South African San Council, for taking the place of Reverend Mario Mahongo at the Stockholm GCC launch event and for doing so brilliantly, despite the shock of his tragic death. Thanks also for her speech at the European Parliament event.

Thanks to Professor Jeffrey Sachs, Dr Leonardo Simão, Dr Mahnaz Vahedi and Vivienne Parry MBE for joining the TRUST team at the European Parliament event.

Thanks to Fritz Schmuhl, the senior editor at Springer, who is still the best book editor I know. This is my sixth Springer book with him, which says it all. Thanks also to George Solomon, the project co-ordinator for this book, for dealing swiftly and efficiently with all questions and for smoothing out any complications in the book production process. Finally, thanks to Ramkumar Rathika for expertly guiding the e-proofing process.

This is also the sixth book for which Paul Wise in South Africa has been the professional copy-editor. Copy-editing sounds like checking that references are in the right format, but that's comparing a mouse to a lion. Paul does a lion's work; he even found a factual mistake in an author biography – written by the author. Thanks, Paul! I hope you're around for the seventh book.

Thanks to Professor Michael Parker, the director of Ethox at Oxford University, for giving a team of us (Joshua Kimani, Leana Snyders, Joyce Adhiambo Odhiambo and me) the floor in his distinguished institute to introduce the GCC.

Thanks to David Coles, Olivia Biernacki, Francesca Cavallaro, Julie Cook, Dieynaba N'Diaye, Francois Bompart, Jacintha Toohey, Rachel Wynberg, Jaci van Niekirk and Myriam Ait Aissa for their contributions to the work, which is summarized in Chapter 8.

Thanks to Julie Cook for brilliant comments on earlier versions of the manuscript, and also to two anonymous peer reviewers for their helpful comments on the book plans.

Thanks to Dr Francesca Cavallaro for creating the educational and fun GCC website.[1]

Thanks to Amy Azra Dean for producing film clips for the GCC website.

---

[1] globalcodeofconduct.org/

Thanks to Kelly Laas, host of the world's largest collection of ethics codes,[2] who has a good answer to *any* question on ethics codes.

Thanks to Julia Dammann for productive Twitter activities on the GCC.

Thanks to Dr Michael Makanga and the European & Developing Countries Clinical Trials Partnership for carrying the costs of the Portuguese translations of the GCC and the San Code of Research Ethics. Thanks to Dr Alexis Holden at the University of Central Lancashire (UCLan), who approved the funding of six further translations.

Thanks to Professor Olga Kubar, Mr Albert Schröder, Dr François Hirsch, Dr Veronique Delpire, Dr Yandong Zhao, Ms Xu Goebel, Dr Shunzo Majima, Dr Dafna Feinholz, Dr Nandini Kumar, Dr Vasantha Muthuswamy, Dr Swapnil S Agarwal, Dr Prabhat K Choudhary, Dr Roli Mathur and Dr Amitabh Dutta for their contributions to the Russian, German, French, Mandarin, Japanese, Spanish and Hindi translations.

Thanks to Robin Richardson, head of the School of Health Sciences at UCLan, for giving Kate Chatfield and me room to do as much work on the GCC as needed, despite other pressing engagements.

Thanks to Denise Bowers, the head of Payroll at UCLan, for supporting an international team at the Centre for Professional Ethics, despite Brexit.

Thanks to Ethan Farrell at UCLan for multiskilled professional administrative support.

Thanks to Geoff Pennington of CD Marketing, Blackburn, for designing the effective GCC brochure and a magnificent document on the origins and history of the San Code of Research Ethics.[3] And thanks too to Clare Danz of the same company for lightning-fast communication and complete reliability.

Thanks to my four co-authors, Dr Kate Chatfield, Dr Michelle Singh, Dr Roger Chennells and Dr Peter Herissone-Kelly. It was a pleasure to work with you.

Last but not least, thanks to Professor Klaus Leisinger for writing the Foreword to this book despite his extremely busy schedule, and for giving the TRUST project and the GCC unprecedented limelight in powerful policy settings.

February 2019   Doris Schroeder

References

EC (2018) A global code of conduct to counter ethics dumping. Infocentre, 27 June. http://ec. europa.eu/research/infocentre/article_en.cfm?id=/research/headlines/news/article_18_06_27_ en.html?infocentre&item=Infocentre&artid=49377

TRUST (2018) Strong speech by Nairobi activist in European Parliament. http://trust-project.eu/ strong-speech-by-nairobi-activist-in-european-parliament/

---

[2] ethicscodescollection.org

[3] http://www.globalcodeofconduct.org/affiliated-codes/

Contents

About the Authors

Doris Schroeder is director of the Centre for Professional Ethics at the University of Central Lancashire, and professor of moral philosophy at the School of Law, UCLan Cyprus. She is the lead author of the Global Code of Conduct for Research in Resource-Poor Settings.

Kate Chatfield is deputy director of the Centre for Professional Ethics, University of Central Lancashire, UK. She is a social science researcher and ethicist specializing in global justice, research ethics, animal ethics and responsible innovation.

Michelle Singh is a project officer at the European & Developing Countries Clinical Trials Partnership in Cape Town, South Africa. She holds a medical PhD and previously managed maternal and child health research studies and clinical trials at the South African Medical Research Council.

Roger Chennells works as legal adviser to the South African San Institute and is a founder-partner in the human rights law practice Chennells Albertyn, Stellenbosch, established in 1981. Specializing in labour, land, environmental and human rights law, he has also worked for Aboriginal people in Australia.

Peter Herissone-Kelly is senior lecturer in philosophy, University of Central Lancashire, UK. He is a specialist in Kantian ethics as well as bioethics, analytic philosophy of language and metaethics.

Abbreviations

ACF	Action contre la Faim
CBD	UN Convention on Biological Diversity
COHRED	Council on Health Research for Development
EC	European Commission (EC)
EDCTP	European and Developing Countries Clinical Trials Partnership
EFPIA	European Federation of Pharmaceutical Industries and Associations
FERCI	Forum for Ethics Review Committees in India
GCC	Global Code of Conduct for Research in Resource-Poor Settings
GVA	Global Values Alliance
HIC	high-income country
Inserm	Institut national de la santé et de la recherche médicale
IPR	intellectual property rights
LMICs	low- and middle-income countries
NGO	nongovernmental organization
PHDA	Partners for Health and Development in Africa
REC	research ethics committee
SASC	South African San Council
SASI	South African San Institute
SWOP	Sex Workers Outreach Programme
UNESCO	United Nations Educational, Scientific and Cultural Organization
WIMSA	Working Group of Indigenous Minorities in Southern Africa

Chapter 1
Ethics Dumping and the Need for a Global Code of Conduct

Abstract The UN's *2030 Agenda for Sustainable Development* calls for more research and innovation to end poverty, leaving no one behind – and yet the export of unethical practices from high-income to lower-income settings is still a major concern. Such ethics dumping occurs in all academic disciplines. When research is regarded, on the one hand, as a dirty word among vulnerable populations who face ethics dumping, and, on the other, as a solution to many of humanity's problems, how can the resulting gulf be bridged? This book describes one initiative to counter ethics dumping: the development and promotion of the Global Code of Conduct for Research in Resource-Poor Settings.

Keywords Ethics dumping · Global research ethics · Exploitation · Vulnerability · Research governance

Research has become a global enterprise. Individual researchers around the world are encouraged to be as mobile as possible (Sugimoto et al. 2017). At the same time, the activities of mobile researchers have made research "one of the dirtiest words in the indigenous world's vocabulary" (Tuhiwai Smith 1999: 1). The indigenous communities in which Tuhiwai Smith, a Māori professor, grew up saw research as something that "told us things already known, suggested things that would not work, and made careers for people who already had jobs" (Tuhiwai Smith 1999: 3).

There is a gulf between those advocating more researcher mobility because "science is the engine of prosperity" (Rodrigues et al. 2016) and those who argue that research can represent harmful "visits by inquisitive and acquisitive strangers" (Tuhiwai Smith 1999: 3). When concerns about ethics dumping[1] are added, this gulf becomes almost unbridgeable.

---

[1] The term was introduced by the Science with and for Society Unit of the European Commission: "Due to the progressive globalisation of research activities, the risk is higher that research with sensitive ethical issues is conducted by European organisations outside the EU in a way that would not be accepted in Europe from an ethical point of view. This exportation of these non-compliant research practices is called ethics dumping" (European Commission nda).

There are two main reasons for ethics dumping – that is, the export of unethical research practices from a high-income to a resource-poor setting. The first is intentional exploitation, where research participants and/or resources in low- and middle-income countries (LMICs) are exploited *on purpose* because the research would be prohibited in the high-income country (HIC). The second is exploitation based on insufficient knowledge or ethics awareness on the part of the mobile researcher. In both cases a lack of adequate oversight mechanisms in the host LMIC is likely to exacerbate the problem (Schroeder et al. 2018).

Examples of ethics dumping in the 21st century include:

- In clinical research, misinterpreting the standard of care, leading to the avoidable deaths of research participants (Srinivasan et al. 2018).
- Research among indigenous populations that led to the publication of "private, pejorative, discriminatory and inappropriate" conclusions and a refusal to engage with indigenous leaders on the informed consent process (Chennells and Steenkamp 2018).
- The export of valuable blood samples from a rural area in China to a US genetic bank, leading to a large amount of research funding for the US team (Zhao and Zhang 2018).
- The use of wild-caught non-human primates in research by a UK researcher who undertook his experiments in Kenya, thus "bypassing British law" (Chatfield and Morton 2018).
- An attempt to seek retrospective ethics approval for a highly sensitive social science study undertaken among vulnerable populations following a local Ebola crisis (Tegli 2018).

How can one reconcile recent cases of ethics dumping with our generation's highly ambitious call for more research and innovation? The United Nations *2030 Agenda for Sustainable Development* aims "to end all forms of poverty… while ensuring that no one is left behind" (UN ndb). To achieve these aims, the UN encourages "fostering innovation" (Goal 9 of Agenda 2030), as "without innovation …, development will not happen" (UN nda).

This book describes one initiative to counter ethics dumping: the development and promotion of the Global Code of Conduct for Research in Resource-Poor Settings (GCC) and its sister code, the San Code of Research Ethics.

The GCC recognizes the considerable power imbalances that may be involved in international collaborative research and provides guidance across all disciplines. It is based on a new ethical framework that is predicated on the values of fairness, respect, care and honesty; values that are imperative for avoiding ethics dumping. The GCC opposes all double standards in research and supports long-term equitable research relationships between partners in lower-income and higher-income settings. This book introduces the GCC in the following manner:

- Chapter 2 reproduces the GCC as launched in the European Parliament in June 2018 and adopted as a mandatory reference document by the European Commission (ndb).

- Chapter 3 explains why values rather than standards, principles, virtues or ideals provide the best guidance in the fight against ethics dumping.
- Chapter 4 answers a philosophical question: how can the GCC can be defended against claims of moral relativism?
- Chapter 5 details 88 risks for ethics dumping, the analytical foundation of the GCC.
- Chapter 6 describes how the GCC was built, from extensive stakeholder engagements to its final translation into Russian, French, Spanish, German, Portuguese, Mandarin, Japanese and Hindi.
- Chapter 7 recounts the history of the San Code of Research Ethics, sister code of the GCC and the first ethics code launched by an indigenous group on the African continent.
- Acknowledging that an ethics code is not enough on its own to counter ethics dumping, Chapter 8 offers advice on community engagement, workable complaints procedures and negotiating fair contracts.
- Chapter 9 presents a brief conclusion.
- The names of the 56 authors of the GCC are set out in the Appendix.

Can an ethics code overcome ethics dumping and bridge the gulf between those for whom international collaborative research is exploitation by strangers, and those who believe it is essential to end all poverty? That is the hope of the authors of the GCC.

References

Chatfield K, Morton D (2018) The use of non-human primates in research. In: Schroeder D, Cook J, Hirsch F, Fenet S, Muthuswamy V (eds) Ethics dumping: case studies from North-South research collaborations, Springer Briefs in Research and Innovation Governance, Berlin, p 81–90

Chennells R, Steenkamp A. (2018) International genomics research involving the San people. In: Schroeder D, Cook J, Hirsch F, Fenet S, Muthuswamy V (eds) Ethics dumping: case studies from North-South research collaborations, Springer Briefs in Research and Innovation Governance, Berlin, p 15–22

European Commission (nda) Horizon 2020: ethics. https://ec.europa.eu/programmes/horizon2020/en/h2020-section/ethics

European Commission (ndb) Participant portal H2020 manual: ethics. http://ec.europa.eu/research/participants/docs/h2020-funding-guide/cross-cutting-issues/ethics_en.htm

Rodrigues ML, Nimrichter L, Cordero RJB (2016) The benefits of scientific mobility and international collaboration. FEMS Microbiology Letters 363(21):fnw247. https://academic.oup.com/femsle/article-lookup/doi/10.1093/femsle/fnw247

Schroeder D, Cook J, Hirsch F, Fenet S, Muthuswamy V (eds) (2018) Ethics dumping: case studies from North-South research collaborations. Springer Briefs in Research and Innovation Governance, Berlin

Srinivasan S, Johari V, Jesani A (2018) Cervical cancer screening in India. In: Schroeder D, Cook J, Hirsch F, Fenet S, Muthuswamy V (eds) Ethics dumping: case studies from North-South research collaborations. Springer Briefs in Research and Innovation Governance, Berlin, p 33–47

Sugimoto CR, Robinson-Garcia N, Murray DS, Yegros-Yegros A, Costas R and Larivière V (2017) Scientists have most impact when they're free to move. Nature 550(7674):29–31

Tegli J (2018) Seeking retrospective approval for a study in resource-constrained Liberia. In: Schroeder D, Cook J, Hirsch F, Fenet S, Muthuswamy V (eds) Ethics dumping: case studies from North-South research collaborations. Springer Briefs in Research and Innovation Governance, Berlin, p 115-119

Tuhiwai Smith, Linda (1999) Decolonizing methodologies: research and indigenous Peoples. Zed Books, London and New York

UN (ndb) The sustainable development agenda. Sustainable Development Goals. https://www.un.org/sustainabledevelopment/development-agenda/

UN (nda) Goal 9. Sustainable Development Goals. https://www.un.org/sustainabledevelopment/infrastructure-industrialization/

Zhao Y, Zhang W (2018) An international collaborative genetic research project conducted in China. In: Schroeder D, Cook J, Hirsch F, Fenet S, Muthuswamy V (eds) Ethics dumping: case studies from North-South research collaborations. Springer Briefs in Research and Innovation Governance, Berlin, p 71–80

Chapter 2
A Value-Based Global Code of Conduct to Counter Ethics Dumping

Abstract The Global Code of Conduct for Research in Resource-Poor Settings (GCC) is designed to counter ethics dumping, i.e. the practice of moving research from a high-income setting to a lower-income setting to circumvent ethical barriers. The GCC is reprinted here. It was completed in May 2018 and adopted by the European Commission as a mandatory reference document for Horizon 2020 in August 2018. For more information on the GCC, please visit: http://www.global-codeofconduct.org/

Keywords Global ethics · Research ethics · International co-operation · Ethics dumping · Low- and middle-income countries

Research partnerships between high-income and lower-income settings can be highly advantageous for both parties. Or they can lead to ethics dumping, the practice of exporting unethical research practices to lower-income settings.

This Global Code of Conduct for Research in Resource-Poor Settings counters ethics dumping by:

Providing guidance across all research disciplines

presenting clear, short statements in simple language to achieve the highest possible accessibility

focusing on research collaborations that entail considerable imbalances of power, resources and knowledge

using a new framework based on the values of fairness, respect, care and honesty

offering a wide range of learning materials and affiliated information to support the Code, and

complementing the European Code of Conduct for Research Integrity through a particular focus on research in resource-poor settings.

Those applying the Code oppose double standards in research and support long-term equitable research relationships between partners in lower-income and high-income settings based on fairness, respect, care and honesty.

Fairness

Article 1

Local relevance of research is essential and should be determined in collaboration with local partners. Research that is not relevant in the location where it is undertaken imposes burdens without benefits.

Article 2

Local communities and research participants should be included throughout the research process, wherever possible, from planning through to post-study feedback and evaluation, to ensure that their perspectives are fairly represented. This approach represents Good Participatory Practice.

Article 3

Feedback about the findings of the research must be given to local communities and research participants. It should be provided in a way that is meaningful, appropriate and readily comprehended.

Article 4

Local researchers should be included, wherever possible, throughout the research process, including in study design, study implementation, data ownership, intellectual property and authorship of publications.

Article 5

Access by researchers to any biological or agricultural resources, human biological materials, traditional knowledge, cultural artefacts or non-renewable resources such as minerals should be subject to the free and prior informed consent of the owners or custodians. Formal agreements should govern the transfer of any material or knowledge to researchers, on terms that are co-developed with resource custodians or knowledge holders.

Article 6

Any research that uses biological materials and associated information such as traditional knowledge or genetic sequence data should clarify to participants the potential monetary and non-monetary benefits that might arise. A culturally appropriate plan to share benefits should be agreed to by all relevant stakeholders, and reviewed regularly as the research evolves. Researchers from high-income settings need to be aware of the power and resource differentials in benefit-sharing discussions, with sustained efforts to bring lower-capacity parties into the dialogue.

Article 7

It is essential to compensate local research support systems, for instance translators, interpreters or local coordinators, fairly for their contribution to research projects.

Respect

Article 8

Potential cultural sensitivities should be explored in advance of research with local communities, research participants and local researchers to avoid violating customary practices. Research is a voluntary exercise for research participants. It is not a mission-driven exercise to impose different ethical values. If researchers from high-income settings cannot agree on a way of undertaking the research that is acceptable to local stakeholders, it should not take place.

Article 9

Community assent should be obtained through recognized local structures, if required locally. While individual consent must not be compromised, assent from the community may be an ethical prerequisite and a sign of respect for the entire community. It is the responsibility of the researcher to find out local requirements.

Article 10

Local ethics review should be sought wherever possible. It is of vital importance that research projects are approved by a research ethics committee in the host country, wherever this exists, even if ethics approval has already been obtained in the high-income setting.

Article 11

Researchers from high-income settings should show respect to host country research ethics committees.

Care

Article 12

Informed consent procedures should be tailored to local requirements to achieve genuine understanding and well-founded decision-making.

Article 13

A clear procedure for feedback, complaints or allegations of misconduct must be offered that gives genuine and appropriate access to all research participants and local partners to express any concerns they may have with the research process. This procedure must be agreed with local partners at the outset of the research.

Article 14

Research that would be severely restricted or prohibited in a high-income setting should not be carried out in a lower-income setting. Exceptions might be permissible in the context of specific local conditions (e.g. diseases not prevalent in high-income countries).

If and when such exceptions are dealt with, the internationally acknowledged compliance commandment "comply or explain" must be used, i.e. exceptions agreed upon by the local stakeholders and researchers must be explicitly and transparently justified and made easily accessible to interested parties.

Article 15

Where research involvement could lead to stigmatization (e.g. research on sexually transmitted diseases), incrimination (e.g. sex work), discrimination or indeterminate personal risk (e.g. research on political beliefs), special measures to ensure the safety and wellbeing of research participants need to be agreed with local partners.

Article 16

Ahead of the research it should be determined whether local resources will be depleted to provide staff or other resources for the new project (e.g. nurses or laboratory staff). If so, the implications should be discussed in detail with local communities, partners and authorities and monitored during the study.

Article 17

In situations where animal welfare regulations are inadequate or non-existent in the local setting compared with the country of origin of the researcher, animal experimentation should always be undertaken in line with the higher standards of protection for animals.

Article 18

In situations where environmental protection and biorisk-related regulations are inadequate or non-existent in the local setting compared with the country of origin of the researcher, research should always be undertaken in line with the higher standards of environmental protection.

Article 19

Where research may involve health, safety or security risks for researchers or expose researchers to conflicts of conscience, tailored risk management plans should be agreed in advance of the research between the research team, local partners and employers.

Honesty

Article 20

A clear understanding should be reached among collaborators with regard to their roles, responsibilities and conduct throughout the research cycle, from study design through to study implementation, review and dissemination. Capacity-building plans for local researchers should be part of these discussions.

Article 21

Lower educational standards, illiteracy or language barriers can never be an excuse for hiding information or providing it incompletely. Information must always be presented honestly and as clearly as possible. Plain language and a non-patronising style in the appropriate local languages should be adopted in communication with research participants who may have difficulties comprehending the research process and requirements.

Article 22

Corruption and bribery of any kind cannot be accepted or supported by researchers from any countries.

Article 23

Lower local data protection standards or compliance procedures can never be an excuse to tolerate the potential for privacy breaches. Special attention must be paid to research participants who are at risk of stigmatization, discrimination or incrimination through the research participation.

Chapter 3
The Four Values Framework: Fairness, Respect, Care and Honesty

Abstract Values inspire, motivate and engage people to discharge obligations or duties. This chapter defends the values approach in the context of guarding against ethics dumping, the practice of exporting unethical research from higher-income to lower-income settings. A number of essential questions will be answered: What are values? What is the meaning of the word "value"? Why does it make sense to choose values as an instrument to guide ethical action in preference to other possibilities? And what is meant by fairness, respect, care and honesty? It is concluded that values can provide excellent guidance and aspiration in the fight against ethics dumping, and are therefore a well-chosen structure for the Global Code of Conduct for Research in Resource-Poor Settings.

Keywords Values · Virtues · Fairness · Respect · Care · Honesty

Introduction

Many celebrated documents which advocate for a better world include a *preamble* that mentions values. For instance, at the international level, the Universal Declaration of Human Rights (UN 1948) lists four values in the first sentence: dignity, freedom, justice and peace in the world. The first sentence of the Convention on Biological Diversity (UN 1992) refers to "the intrinsic *value* of biological diversity and of the ecological, genetic, social, economic, scientific, educational, cultural, recreational and aesthetic *values* of biological diversity" (emphasis added).

Other national or professional codes have incorporated values prominently *into* individual articles. For instance, at the national level in the UK, the first item of *The Code: Professional Standards of Practice and Behaviour for Nurses, Midwives and Nursing Associates*, reads: "Treat people with kindness, respect and compassion" (NMC 2018).

In some codes one has to search to find obvious references to values as they are often incorporated in a more implicit manner, such as in the Declaration of Helsinki

(WMA 2013), which speaks of "safety, effectiveness, efficiency, accessibility and quality" in article 6.

When developing the Global Code of Conduct for Research in Resource-Poor Settings (GCC), a unique approach emerged naturally from the process employed. Its underpinning values materialized *ahead* of its final articles through an investigation into the risks of exploitation in international collaborative research (Chapter 5), and from a global engagement and fact-finding mission (Chapter 6).

It soon became clear that fairness, respect, care and honesty are all lacking, or deficient, whenever ethics dumping[1] occurs, and that a loss of trust in researchers and research itself can result. What also emerged is that these values are shared across the range of cultures that were represented in the TRUST[2] consortium. It was therefore possible to surmise that these shared values are vital for equitable research partnerships and to prevent ethics dumping. In other words, these values are necessary to foster an ethical culture in research, and are therefore values to which all researchers should aspire.

This chapter will answer some essential questions: What are values? What is the meaning of the word "value"? Why does it make sense to choose values as an instrument to guide ethical action in preference to other possibilities? And finally, what is meant by "fairness", "respect", "care" and "honesty"?

The Meaning of "Value"

Values pervade human experience (Ogletree 2004), and references to "values" are ubiquitous. With vast numbers of articles, books and internet sites offering advice on matters such as *values we should live by*, *discovering our own values*, *changing our core values* and *achieving success through values*, it is obvious that values are important to people.

The term "value" can be used in many different ways.[3] With reference to the way in which people use the term, three primary meanings of "value" can be distinguished (see Fig. 3.1).

First, value can refer to measurability. Mathematics operates with values, which can, for example, be discrete or continuous. Artists might speak of colours having values, meaning the relative lightness or darkness of a colour. In music, a note value determines the duration of a musical note. Economists or art dealers might measure value in monetary terms; a particular company or a particular painting might be

---

[1] The export of unethical research from a high-income setting to a resource-poor setting with weaker compliance structures or legal governance mechanisms.

[2] TRUST was an EU-funded project which operated from 2015 to 2018 and developed the GCC, among other outputs. http://trust-project.eu/

[3] This section draws on unpublished work by Professor Michael Davis, a philosopher specializing in professional ethics.

Fig. 3.1 The meaning of value

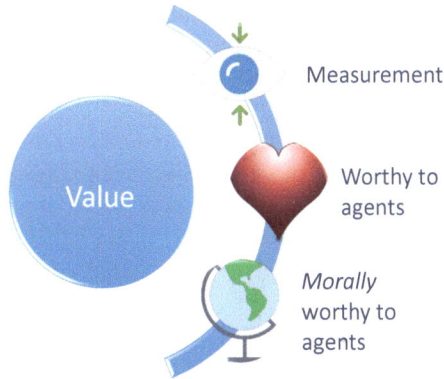

valued at a certain amount of money. Value, in this sense of the word, has no relationship to values such as admiration, approval or motivation.

Secondly, people can value certain features or entities. For instance, somebody might value money, fame or glory. For value to exist, there must be an agent (a person) who is doing the valuing, and the feature or entity must be worth something to this agent (Klein 2017). The values of one individual can be very different from those of another person. For instance, a regular income is worth a lot to a person who values routine and security; it can contribute to their wellbeing and happiness. Others, who value personal freedom more than routine and security, might be just as happy with occasional income, as long as they are not bound to a nine-to-five job. If most humans around the world value a particular thing, it can be described as a universal value.

Thirdly, values can refer to goals and ambitions, with a moral connotation. In business literature, for example, one often finds reference to value-led management or organizational values, and many institutions make a point of establishing, promoting and broadcasting their values. For instance, the stated values of the University of Central Lancashire (UCLan), at which several of the authors of this book are based, are: common sense, compassion, teamwork, attention to detail and trust (UCLan nd). These values are all morally positive and they are intended to guide the actions of students, staff and the institution itself. In this third sense of the word, moral values "will enable us to determine what is morally right or what is valuable in particular circumstances" (Raz 2001: 208). If most humans around the world share a particular moral value, it can be described as a universal moral value.

There are numerous advantages to having credible moral values at the level of organizations. Such values influence the culture of an organization (Martins and Coetzee 2011), which in turn has a positive impact upon corporate performance (Ofori and Sokro 2010), and job stress and satisfaction (Mansor and Tayib 2010), as well as business performance and competitive advantage (Crabb 2011). Furthermore, when employees' values are aligned with organizational values, this benefits both the wellbeing of individuals and the success of the organization (Posner 2010).

There are many internet sites that offer lists of core values. One of them (Threads Culture nd) includes 500 values, from "above and beyond" to "work life balance".

Not all of these are moral values. For instance, this particular list includes values such as clean, exuberant, hygienic, neat, poised and winning (Threads Culture nd). Another site lists 50 values, including authenticity, loyalty and wisdom, and advises that fewer than five should be selected for leadership purposes (Clear nd).

The GCC is structured around four moral values: fairness, respect, care and honesty. These four values were not chosen from any existing lists; they emerged through in-depth consultation efforts around the globe (chapter 6). But why did the TRUST team choose moral values rather than other action-guiding moral modes for the GCC?

What Can Guide Moral Action?

The GCC is based on moral values, but the code authors could have opted to frame the code and guide action in other ways, including the following:

- *Standards* is a technical term used to achieve desired action. Standards are precise and give exact specifications, which are in many cases measurable, as in the maximum vehicle emissions allowed for cars. Standards can also be used in ethics. For instance, a well-known voluntary standard to guide ethical action is ISO 26000 (ISO not dated), developed by the International Organization for Standardization. ISO 26000 assesses the social responsibility of companies. Its guidance includes prohibitions against bribery, and the requirement to be accountable for any environmental damage caused.
- *Principles* are behavioural rules for concrete action. When you know the principle, you know what to do. For instance, *in dubio pro reo* has saved many innocent people from going to jail as it gives the courts very concrete advice. Literally translated, it means, "when in doubt, then for the accused" (a person remains innocent until proven guilty). This principle goes back to both Aristotle and Roman law.
- *Virtues* are beneficial character traits that human beings need to flourish (Foot 1978: 2f). One can observe them in real people or in fictional characters. England's semimythical Robin Hood, for instance, is seen as courageous and benevolent. He fights a David-and-Goliath battle against the Sheriff of Nottingham (courage) so that the poor have food (benevolence). Like values, for virtues to exist, there must be an agent (a person) who is being virtuous; virtues focus on the moral agent rather than on the standard or principle that underlies a decision.
- *Ideals* drive towards perfection and are highly aspirational. Some people will say "in an ideal world" to denote that something is unrealistic from the start. The ancient Greek philosopher Aristotle argued that we should strive towards perfection of character and that ideals can be guiding lights in character building. "Good character is an ideal outside of oneself that all strive for" (Mitchell 2015).

So why were values chosen as the foundation for the GCC rather than standards, principles, virtues or ideals?

Ideals are the most aspirational of the concepts available to guide ethical action. However, hardly anybody can live up to all of their ideals. If one phrased an ethics code around ideals, those who should be led by the code might suggest that not reaching the ideals on every occasion would be acceptable. This is not the case. The 23 articles of the GCC (chapter 2) are not aspirational. They are mandatory.

Virtues are found both historically and internationally in many important documents of learning and wisdom. Famously, Aristotle (384–322 BC) linked human "happiness and wellbeing" to "leading an ethical life", guided by the cardinal values of courage, justice, modesty and wisdom (Aristotle 2004). According to Confucianism, the most important traditional virtues are said to be benevolence, righteousness, propriety, wisdom, trustworthiness, filial piety, loyalty and reciprocity (Wang et al. 2018). Virtues are a good way to drive ethical action, in particular global ethical action, but the TRUST team had good reason not to use virtues as the foundation of the GCC.

Virtues can be regarded as *embodied* ethical values because they are manifested in persons. One can learn a lot by observing real people (such as Mother Theresa or Nelson Mandela) and following their example. This makes virtue approaches very useful in leadership and mentoring (Resnik 2012). But not every researcher has access to mentors and learning via example. Besides, early career researchers are said to benefit more from rule-based approaches (Resnik 2012). Hence, while virtues were considered as a possibility for the foundation of the GCC, they were excluded because of their strong reliance upon the availability of role models.

Principles have a long-standing tradition in practical moral frameworks, especially principlism, the moral framework relating to bioethics developed by Beauchamp and Childress (2013). As argued in Chapter 4, we believe that the four principles of Beauchamp and Childress – autonomy, non-maleficence (do no harm), beneficence and justice – should instead be called values. Principles, as we understand them, are more concrete than values. Principles can provide almost immediate and very straightforward answers to ethical questions.

A famous principle in political philosophy is Rawls's difference principle. The principle holds that divergence from an egalitarian distribution of social goods (e.g. income, wealth, power) is only allowed when this non-egalitarian distribution favours the least advantaged in society (Rawls 1999: 65–70). In other words, if a particularly talented wealth creator increases the overall wealth pie so that the least advantaged in society are better off, she can receive a bigger share of the pie than others. Knowing about this principle gives answers to social philosophy questions, which the value of fairness or justice would not. Rawls *applied* the value of fairness to derive the more concrete difference principle. Principles are therefore too concrete and too prescriptive to form the foundation of the GCC. They would not leave enough room for local agreements between partners from high- and lower-income settings as envisaged by various GCC articles, such as article 1: "Local relevance of research … should be determined in collaboration with local partners."

Standards are even more specific than principles and have an even stronger action-guiding function. They prescribe very concrete activities in given settings. To formulate standards for ethical interaction between partners from different settings would certainly be too prescriptive. A standard cannot be diverged from (for example, a limit to vehicle emissions). For instance, if article 10[4] were a standard, no exception to double ethics review would be possible. But there may be good reason to allow such an exception in certain circumstances. For instance, if ethics approval has been given in a high-income setting and community approval obtained in a host setting where no ethics committee operates, then it may be perfectly ethical to proceed.

The San community in South Africa, for instance, has no facility for providing ethics committee approval, but the South African San Council can provide community approval for research projects in the community (Chapter 7). A *standard* of double ethics review would forbid any research in the San community until an ethics committee were established, which might even undermine the San people's self-determined research governance structures. For this reason, it is clear that standards are too prescriptive to be applied to every setting, and might hinder valuable research.

This leaves ethical values, which operate as guides on the route to doing the right thing and are not overly prescriptive. They do not undermine the need to develop bespoke agreements across cultures via discussions between research teams and communities. At the same time, there is another, positive reason to choose values as the foundation for the GCC. Values inspire and motivate people to take action – and that is exactly what is needed to guard against ethics dumping.

Values and Their Motivating Power

Research stakeholders who are guided by values will hopefully be inspired and motivated by the GCC and not just follow its rules reluctantly or grudgingly. Why is that? Values can serve as motivating factors in promoting or inhibiting human action (Marcum 2008, Locke 1991, Ogletree 2004). The influence of personal values upon behaviour has become a subject of extensive research in the social sciences and in psychology, particularly over the past forty years, with just about every area of life being examined through the lens of personal values – for example, consumer practices (Pinto et al. 2011), political voting habits (Kaufmann 2016), employee creativity (Sousa and Coelho 2011), healthcare decisions (Huijer and Van Leeuwen 2000), investment decisions (Pasewark and Riley 2010), and sexuality and disability (Wolfe 1997), to name but a few.

---

[4] Local ethics review should be sought wherever possible. It is of vital importance that research projects are approved by a research ethics committee in the host country, wherever this exists, even if ethics approval has already been obtained in the high-income setting.

Arguably the most prominent theory of the motivational power of human values was developed by social psychologist Shalom Schwartz, back in 1992. Schwartz's theory of basic values is distinctive because, unlike most other theories, it has been tested via extensive empirical investigation. Studies undertaken since the early 1990s have generated large data sets from 82 countries, including highly diverse geographic, cultural, religious, age and occupational groups (Schwartz 2012). Findings from Schwartz's global studies indicate that values are inextricably linked to affect. He claims that when values are activated, they become infused with feeling (Schwartz 2012). For example, people for whom routine and security are important values will become disturbed when their employment is threatened and may fall into despair if they actually lose their jobs. Correspondingly, when moral values like fairness or respect are important, people will react when they witness instances of unfairness or disrespect; they will feel motivated to respond in some way.

Schwartz's research investigated motivational values in general (combining our second and third meanings of "value"), and not just moral values. As noted earlier, people can be motivated by many different values, but interestingly, when asked to rank values in order of importance, the participants in Schwartz's studies consistently rated those with explicit moral connotations as the most important values (Schwartz 2012). This suggests that people hold their moral values in high esteem and can be strongly influenced by them.

From Values to Action

Ethical values give us direction but are not sufficient to make us ethical researchers who avoid ethics dumping. One can hold the value of honesty and yet fail to be an honest person. One can hold the value of respect and yet cause harm when disrespecting local customs. Values can motivate and they can help to establish moral goals, but they do not explain how to achieve them. A means of operationalizing values is needed.

One method would be to cultivate virtues that are aligned with the values. As noted above, virtues are positive character traits individuals build over time which are needed for human flourishing. Once a value such as honesty becomes second nature, one can say that honesty is a virtue of that person. If all researchers developed the virtues of fairness, respect, care and honesty, then being an ethical researcher would come naturally to them. However, this is far from easy, and the development of virtues takes time. It is perhaps possible for researchers who have worked in the field for many years, and have a wealth of knowledge and experience, but certainly not for young researchers who need training, guidance and practice.

Daniel Russell (2015: 37f) illustrates the challenge for virtue ethics in guiding specific action when he asks us to think about generosity:

> Sometimes helping means giving a little, sometimes it means giving a lot; sometimes it means giving money, sometimes it means giving time, or just a sympathetic ear; sometimes it means offering advice, sometimes it means minding one's own business; and which of

these it might mean in this case will depend on such different things as my relationship with my friend, what I am actually able to offer, why and how often my friend has problems of this kind, and so on.

For all those who are still developing their virtues, a code such as the GCC can help to guide action. As noted at the outset, people are much more contented and productive when their own values are aligned with company or institutional values and rules. It therefore made sense to align the articles of the GCC with those values that are necessary for ethical research and to which researchers must aspire. The values of fairness, respect, care and honesty provide the ethos, the motivation and the goals for ethical research. The 23 articles making up the GCC therefore enable operationalization of the values.

This leaves the task of outlining what is meant by each of the four values of fairness, respect, care and honesty, keeping in mind the following important points. First, precise specifications of values might be affected by customs and preferences, so that different cultures have different views on the *exact* content of the values. Second, the importance of *process* cannot be underestimated. The reason why articles 2[5] and 4[6] of the GCC emphasize inclusion is that the specification of what each value requires in a given setting needs to be determined collaboratively. As a result, this sketch of the content of the four values is brief and leaves room for regional variations.

The Four Values

Fairness

The terms "fairness", "justice" and "equity" are often used interchangeably. The TRUST consortium chose the term "fairness" in the belief that it would be the most widely understood globally. Philosophers commonly distinguish between four types of fairness (Pogge 2006) (see Fig. 3.2).

The most relevant fairness concepts in global research ethics are fairness in exchange and corrective fairness. In global collaborations, at least two parties are involved in a range of transactions. Typical fairness issues between partners from high-income countries (HICs) and those from low- and middle-income countries (LMICs) are:

• Is the research relevant to local research needs?
• Will benefit sharing take place?
• Are authors from LMICs involved in publications?

---

[5] Local communities and research participants should be included throughout the research process, wherever possible, from planning through to post-study feedback and evaluation, to ensure that their perspectives are fairly represented. This approach represents Good Participatory Practice.

[6] Local researchers should be included, wherever possible, throughout the research process, including in study design, study implementation, data ownership, intellectual property and authorship of publications.

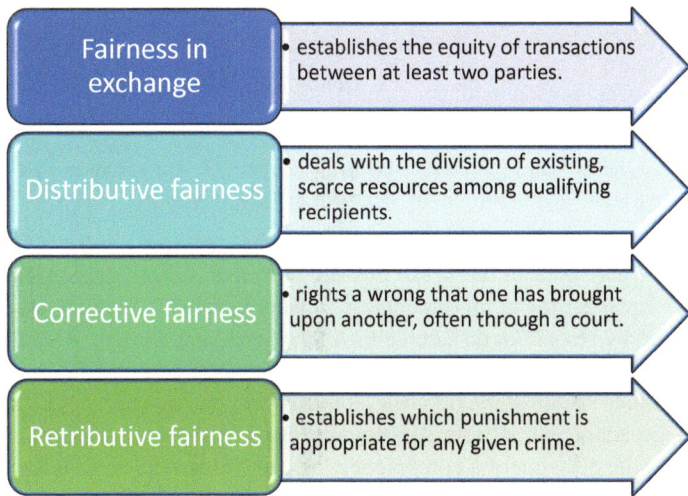

Fig. 3.2 Types of fairness

These are questions about *fairness in exchange*. For instance, LMIC research participants contribute to the progress of science, but this is only fair if the research is relevant to their own community or if other benefits are received where this is not possible. For instance, to carry the burden of a clinical study is only worthwhile for a community if the disease under investigation occurs locally and the end product will become available locally.

Corrective fairness, which presupposes the availability of legal instruments and access to mechanisms to right a wrong (e.g. a complaints procedure, a court, an ethics committee) is also important in global research collaborations. For instance, if no host country research ethics structure exists, corrective fairness is limited to the research ethics structure in the HIC, which may not have the capacity to make culturally sensitive decisions.

The broader question of what HICs owe LMICs falls under *distributive fairness*. One can illustrate the difference between fairness in exchange and distributive fairness using the example of post-study access to successfully tested drugs. In the first case (fairness in exchange) one could argue that research participants have contributed to the marketing of a particular drug and are therefore owed post-study access to it (should they need the drug to promote their health and wellbeing, and should they not otherwise have access to it). In the second case (distributive fairness) one could provide a range of arguments, for instance being a signatory to the Universal Declaration of Human Rights (UN 1948), to maintain that *all* human beings who need the drug should have access to it, and not just the research participants. These wider fairness issues cannot be resolved by researchers and are therefore not directly included in the GCC. Likewise, *retributive fairness* is less relevant as few ethics violations fall under the punitive and criminal law, and if they do, it is indeed criminal law that should be used to deal with a fairness violation.

Respect

The term "respect" is used in many ethics frameworks. For instance, the Declaration of Helsinki (WMA 2013) notes in article 7:

> Medical research is subject to ethical standards that promote and ensure *respect* for all human subjects and protect their health and rights. (emphasis added)

Its ubiquitous use does not, however, mean that "respect" is a clear term. In everyday life, it is used in the sense of deep admiration. For instance, somebody could say, "I respect the achievements of Nelson Mandela". However, that is not what is meant by respect in research ethics. The statement from the Declaration of Helsinki does not mean that research participants must be admired. To be respected in research ethics is almost the opposite. It means that one must accept a decision or a way of approaching a matter, even if one disagrees strongly. A case in point would be respecting the decision of a competent adult Jehovah's Witness to refuse a blood transfusion for reasons of religious belief, even if this means certain death.

Respect is therefore a difficult value, as there will be cases where one *cannot* accept another's decision. For instance, if a researcher learns about female genital mutilation being used as a "cure" for diarrhoea in female babies (Luc and Altare 2018), respecting this approach to health care is likely to be the wrong decision – particularly as the practice is probably illegal. But the fact that respect may be difficult to operationalize in global research collaborations does not mean that it is a value one can dispense with.

There are many possible ways of showing respect that do not create conflicts of conscience. For instance, illiterate San community members should not be enrolled in research studies unless San leaders have been contacted first, in accordance with community systems. And researchers from HICs should not insist that LMIC ethics committees accept the format of the researchers' preferred ethics approval submission; instead the HIC researchers should submit the study for approval in the format required by the LMIC committee. This shows respect in international collaborative research.

While it may be difficult to imagine a situation where an HIC researcher is accused of being too fair, too honest or too caring, it is possible to be accused of being "too respectful" – for instance, if one tolerates major violations of human rights. It is indeed sometimes difficult to strike a balance between dogmatically imposing one's own approach and carelessly accepting human rights violations, but that is the balance researchers should strive for.

Care

Sometimes one word describes different concepts. This is the case with "care". The statement, "I care for my grandfather," can mean two diametrically opposed things. First, it could mean that the person is very attached to her grandfather even though she hardly ever sees him. Second, it could mean that she is the person who injects her grandfather with insulin, cooks his meals, and makes sure that his needs are taken care of every day, even if there is antipathy between them.

The meaning of the value of care in the context of global research ethics links more to the second use of the term; to look after or take care of somebody or something. As a main priority, one should take care of the interests of those enrolled in research studies to the extent that one always prioritizes their welfare over any other goals – for example, accepting the decisions of those who choose to withdraw from an ongoing study, even if this impairs the project's results. In line with article 8 of the Declaration of Helsinki (WMA 2013) that means:

> While the primary purpose of medical research is to generate new knowledge, this goal can never take precedence over the rights and interests of individual research subjects.

This care applies across disciplines, not only in medical research, and it is not restricted to human research participants. Article 21 of the Declaration of Helsinki (WMA 2013) extends the care for research subjects' welfare to research animals. Likewise, care for environmental protection is increasingly included in research ethics processes and frameworks for responsible research. For instance, the European Commission's Horizon 2020 ethics review process addresses potentially negative impacts on the environment (Directorate General for Research 2019: section 7). Richard Owen et al. (2013) define responsible research and innovation as "a collective commitment of care for the future through responsive stewardship of science and innovation in the present", a statement that has clear relevance to environmental protection.

Researchers who take care to avoid negative impacts in their work will not "helicopter" in and out of a research area they are not familiar with, but will use systems of *due diligence* to ensure that risks are assessed and mitigated. For instance, an HIC research team that strips a local area of all doctors and nurses by attracting them into their high-tech research facility is not acting carefully and ethically.

Ideally, researchers who take good care will combine the two concepts mentioned above: they care about research participants, in the sense that the participants are important to them, *and* they feel responsible for the welfare and interests of those who contribute to their research, or might suffer as a result of it (including animals and the environment).

Honesty

Honesty is a value that does not need complicated explanations or definitions. In all cultures and nations, "Do not lie" is a basic prerequisite for ethical human interaction. It is so basic a value that its synonyms are often broad ethics terms. For instance, according to Google (2018), synonyms for "honesty" are:

> moral correctness, uprightness, honourableness, honour, integrity, morals, morality, ethics, principle, (high) principles, nobility, righteousness, rectitude, right-mindedness, upstandingness

What does need explaining, however, is the scope of the value of honesty in the context of global research ethics. Telling lies is only one possible wrongdoing in the context of a broad understanding of honesty. For instance, in research ethics it is equally unacceptable to leave out salient features from an informed consent process. While this might, strictly speaking, not involve a lie, concealing important information that might make a difference to someone's consent violates the value of honesty as much as lying. For this reason, research ethicists often use the terms "transparency" and "open communication" to ensure that all relevant information is provided so that research participants can make an informed choice about whether to participate or not.

In addition to lying and withholding information, there are other ways of being dishonest, in the sense of not communicating openly and transparently. For instance, in a vulnerable population with high levels of illiteracy, it can be predicted that a printed information sheet about research will not achieve *informed* consent. The same can be said for a conscious failure to overcome language barriers in a meaningful way: leaving highly technical English terms untranslated in information sheets can easily lead to misunderstandings.

Honesty is also related to research conduct other than interaction with research participants. Most prominently, the duties of honesty are described in *research integrity* frameworks: do not manipulate your data, do not put your name onto publications to which you have not contributed, do not waste research funds, to give only three examples. However, while the latter prescriptions for conduct with integrity in research are important, they are not directly linked to exploitation in global research collaboration and are not covered in the GCC. In this context, the European Code of Conduct for Research Integrity (ALLEA, 2017) is very helpful.

Conclusion

Standards, principles, values, virtues *and* ideals can guide moral action. At the foundation of the GCC are values. Why? For three main reasons:

1. Values inspire action; they motivate people to do things. For instance, when the value of fairness is threatened, people normally respond with action.
2. Values provide the golden middle way between being overly prescriptive and overly aspirational. Standards and principles require too much precision in their formulation and are too prescriptive in international collaborative research, while virtues and ideals are too aspirational in their demands of researchers.
3. Values emerged naturally from the major engagement activities undertaken prior to developing the GCC.

The eradication of ethics dumping requires not only moral *guidance* but also moral *action* to counter violations of fairness, respect, care and honesty. The 23 short, accessible articles of the GCC are intended to both guide *and* inspire researchers to act with fairness, respect, care and honesty.

References

ALLEA (2017) The European code of conduct for research integrity. All European Academies, Berlin. https://www.allea.org/allea-publishes-revised-edition-european-code-conduct-research-integrity/

Aristotle (2004) Nicomachean Ethics (trans: Thomson JAK), 2nd edn. Tredennick H (ed). Penguin Classics, London

Beauchamp TL, Childress JF (2013) Principles of biomedical ethics, 7th edn. Oxford University Press, New York

Clear J (nd) Core values list. https://jamesclear.com/core-values

Crabb S (2011) The use of coaching principles to foster employee engagement. The Coaching Psychologist 7(1):27–34

Directorate General for Research (2019) Horizon 2020 Programme: guidance – how to complete your ethics self-assessment, version 6.1. European Commission. http://ec.europa.eu/research/participants/data/ref/h2020/grants_manual/hi/ethics/h2020_hi_ethics-self-assess_en.pdf

Google (nd) Honesty: synonyms. https://www.google.com/search?q=honesty&ie=&oe=

Google (2018) Google search for "Honesty" conducted on 24 November 2018.

Huijer M, van Leeuwen E (2000) Personal values and cancer treatment refusal. Journal of Medical Ethics 26(5):358–362

ISO (nd) ISO 26000: social responsibility. International Organization for Standardization. https://www.iso.org/iso-26000-social-responsibility.html

Kaufmann E (2016) It's NOT the economy, stupid: Brexit as a story of personal values. British Politics and Policy. London School of Economics and Political Science. http://blogs.lse.ac.uk/politicsandpolicy/personal-values-brexit-vote/

Klein LA (2017) A free press is necessary for a strong democracy. ABA Journal. http://www.abajournal.com/magazine/article/free_press_linda_klein?icn=most_read

Locke EA (1991) The motivation sequence, the motivation hub, and the motivation core. Organizational Behavior and Human Decision Processes 50(2):288–299

Luc G, Altare C (2018) Social science research in a humanitarian emergency context. In: Schroeder D, Cook J, Hirsch F, Fenet S, Muthuswamy V (eds) Ethics dumping: case studies from North-South research collaborations. Springer Briefs in Research and Innovation Governance, Berlin, p 9–14. https://link.springer.com/book/10.1007%2F978-3-319-64731-9

Mansor M, Tayib D (2010). An empirical examination of organisational culture, job stress and job satisfaction within the indirect tax administration in Malaysia. International Journal of Business and Social Science 1(1):81–95

Marcum JA (2008). Medical axiology and values. In: An introductory philosophy of medicine: humanizing modern medicine. Philosophy and Medicine 99. Springer Science and Business Media, p 189–205

Martins N, Coetzee M (2011) Staff perceptions of organisational values in a large South African manufacturing company: exploring socio-demographic differences. SA Journal of Industrial Psychology 37(1):1–11

Mitchell LA (2015) Integrity and virtue: the forming of good character. The Linacre Quarterly 82(2):149–169

NMC (2018) The code: professional standards of practice and behaviour for nurses, midwives and nursing associates. Nursing & Midwifery Council. https://www.nmc.org.uk/standards/code/

Ofori DF, Sokro E (2010). Examining the impact of organisational values on corporate performance in selected Ghanaian companies. Global Management Journal 2(1)

Ogletree TW (2004) Value and valuation. In: Post SG (ed) Encyclopedia of bioethics, 3rd edn. MacMillan Reference USA, New York, p 2539–2545

Owen R, Stilgoe J, Macnaghten P, Gorman M, Fisher E, Guston D (2013) A framework for responsible innovation. In: Owen, R, Bessant J, Heintz M (eds) Responsible innovation: managing the responsible emergence of science and innovation in society. John Wiley, London, p 27–50

Pasewark WR, Riley ME (2010) It's a matter of principle: the role of personal values in investment decisions. Journal of Business Ethics 93(2):237–253

Philippa Foot P (1978) Virtues and vices. In: Virtues and vices and other essays in moral philosophy. Blackwell, Oxford, p 1–18

Pinto DC, Nique WM, Añaña EDS, Herter MM (2011) Green consumer values: how do personal values influence environmentally responsible water consumption? International Journal of Consumer Studies 35(2):122–131

Pogge T (2006) Justice. In: Borchert DM (ed) Encyclopedia of philosophy, 2nd edn (vol 4). Macmillan Reference, Detroit MI, p 862–870

Posner BZ (2010) Another look at the impact of personal and organizational values congruency. Journal of Business Ethics 97(4):535–541

Rawls J (1999) A theory of justice, revised edn. Oxford University Press, Oxford

Raz J (2001) Engaging reason: On the theory of value and action. Oxford University Press, Oxford

Resnik DB (2012) Ethical virtues in scientific research. Accountability in Research 19(6):329–343

Russell D C (2015) Aristotle on cultivating virtue. In: Snow, N (ed) Cultivating virtue: perspectives from philosophy, theology, and psychology. Oxford University Press, Oxford, p 37–38

Schwartz SH (2012) An overview of the Schwartz theory of basic values. Online Readings in Psychology and Culture 2(1):11

Sousa CM, Coelho F (2011) From personal values to creativity: evidence from frontline service employees. European Journal of Marketing 45(7/8):1029–1050

Threads Culture (nd) Core values examples. https://www.threadsculture.com/core-values-examples/

UCLan (nd) The UCLan values. University of Central Lancashire. https://www.uclan.ac.uk/work/life-at-uclan.php

UN (1948) Universal Declaration of Human Rights. http://www.un.org/en/universal-declaration-human-rights/

UN (1992) Convention on Biological Diversity. https://www.cbd.int/doc/legal/cbd-en.pdf

Wang X, Li F, Sun Q (2018) Confucian ethics, moral foundations, and shareholder value perspectives: an exploratory study. Business Ethics: A European Review 27(3):260–271

WMA (2013) Declaration of Helsinki. World Medical Association. https://www.wma.net/policies-post/wma-declaration-of-helsinki-ethical-principles-for-medical-research-involving-human-subjects/

Wolfe PS (1997) The influence of personal values on issues of sexuality and disability. Sexuality and disability 15(2):69–90

Chapter 4
Respect and a *Global* Code of Conduct?

Abstract The Global Code of Conduct for Research in Resource-Poor Settings claims *global* applicability and promotes respect as one of its four values. Hence, the code anticipates potentially unresolvable differences between cultures, while maintaining it is globally valid. Examining, but discarding, several possibilities to deal with normative relativism, this chapter argues, with Beauchamp and Childress (2013, *Principles of Biomedical Ethics*, 7th edn. Oxford University Press, New York) that values can be internal to morality itself, allowing their global applicability.

Keywords Values · Normative relativism · Global justice · Research ethics · Fairness · Principlism

Introduction

The Global Code of Conduct for Research in Resource-Poor Settings (GCC) is built around four values: fairness, respect, care and honesty. In this chapter, we tackle the moral relativism claim against values approaches. Some readers may feel that no such effort is necessary. It may, in their view, be obvious that fairness, respect, care and honesty are worthy values. They may also believe that these values have application in global research ethics and that they can counter ethics dumping, the practice of moving unethical research from a high-income setting to a resource-poor setting, which – by definition – requires a global approach. For those who are more sceptical, we sketch a plausible response to the moral relativism objection.

The GCC's value of respect recognizes significant variation in cultural norms and practices (for example in article 8[1]), while implicitly assuming that the four

---

[1] Potential cultural sensitivities should be explored in advance of research with local communities, research participants and local researchers to avoid violating customary practices. Research is a

© The Author(s), under exclusive license to Springer Nature Switzerland AG 2019   27
D. Schroeder et al., *Equitable Research Partnerships*, SpringerBriefs in Research and Innovation Governance, https://doi.org/10.1007/978-3-030-15745-6_4

values the code recommends are globally applicable. How do we reconcile this tension? That is, how do we demonstrate that, in making use of the four values, one group is not illegitimately imposing its values on others, in ways that the GCC itself would deem unacceptable?

The surest way of doing this is to defend the claim that the four values, which we believe have particular application in research in resource-poor settings, could be *global* or universal values. Let us call this claim "the global applicability thesis". We need somehow, then, to be able to maintain the global applicability thesis alongside the recognition of significant variation in norms across cultures. We will look in turn, in the sections that follow, at a number of suggestions about how this may be achieved.

First we will consider the possibility that the requirement to proceed with fairness, respect, care and honesty leads to an acceptance of a thoroughgoing moral relativism – that is, a robust and unflinching commitment to the belief that all values are culture-bound, and that there are no "extra-cultural" values or norms. We will argue that, for reasons articulated by Bernard Williams (1972) nearly half a century ago, such strict moral relativism is unsustainable.

Then we will consider the merits of a more moderate moral relativism, of the sort argued for by the Chinese-American philosopher David Wong (1991, 2009). This approach combines a recognition of variation in norms across cultures with a certain sort of universalism. For Wong, what remains constant across disparate systems of moral norms is the *purpose* behind any such system, or the aim of morality as such. We will argue that Wong's approach, though an improvement on a more extreme relativism, does not provide what we need: that is, it does not show there to be some universal norms – among which are our four values – in addition to some genuine cross-cultural normative variation.

Finally we introduce an approach that has much in common with the TRUST[2] approach: the "four-principles approach" presented by Tom Beauchamp and James Childress in successive editions of their book *Principles of Biomedical Ethics* (2013). Beauchamp and Childress maintain that there are four central values/principles[3] (see the box below for the difference between the two) that are especially applicable to their own area of ethical interest, biomedical ethics. They use the term *principles*, and identify them as respect for autonomy, non-maleficence (do no harm), beneficence and justice (Beauchamp and Childress 2013).

---

voluntary exercise for research participants. It is not a mission-driven exercise to impose different ethical values. If researchers from high-income settings cannot agree on a way of undertaking the research that is acceptable to local stakeholders, it should not take place.

[2] EU-funded research project, which developed the GCC from 2015 to 2018.

[3] According to the definition of values in the box (and in Chapter 3 of this book), the four-principles approach should be called the four-values approach, but this makes no difference in substance.

Values and Principles

The words "values" and "principles" are often used interchangeably. We will distinguish them as below.

If people *value* something, they hold it dear, and they believe it is of high importance. This could be power, money or kindness; values are not necessarily morally positive. *Ethical* values, on the other hand, are guides on the route to doing the right thing or developing a moral character. They are by definition morally positive. For example, greed is not an ethical value, but generosity is.

A *principle* is a behavioural rule for concrete action. When you know the principle, you know what to do. For instance, the principle *in dubio pro reo* has saved many innocent people from going to jail as it gives courts very concrete advice. It means, "When in doubt, then favour the accused," (in other words, "innocent until proven guilty") and goes back to both Aristotle and Roman law.

Beauchamp and Childress maintain that their four principles are globally applicable – that is to say, they are universally relevant to the sorts of ethical questions that arise in biomedicine; they are every bit as integral to the understanding and resolution of medical ethics problems in Bangkok as they are in Boston, equally pertinent in both Cape Town and Copenhagen. Their status as globally applicable is, according to Beauchamp and Childress, underwritten by their forming part of what they call "the common morality", understood as a system of general norms that will be specified differently in different cultures, but to which all morally committed persons everywhere will subscribe.

We want to argue that the four values – fairness, respect, care and honesty – are rooted in a globally applicable common morality, the norms of which can be specified in various ways in disparate cultures. Insofar as this is the case, our argumentational strategy will be similar to that of Beauchamp and Childress.

However, we want to maintain that the four GCC values, taken together, have a less contentious claim to be globally applicable than Beauchamp and Childress's principles, for two reasons.

1. One of Beauchamp and Childress's principles – respect for autonomy – has often, and with some justification, been criticized for being culturally bound rather than universal (Huxtable 2013, Kara 2007, Cheng-Tek Tai 2013, Kiak Min 2017)
2. The GCC's four-values approach was developed collaboratively with diverse stakeholders *from all continents*, including significant representation from vulnerable research populations (see Chapters 6 and 7).

The Four Values and Moral Relativism

Giving the value of respect high standing, as one of only four values in the GCC framework, opens the framework to attack for being morally relative. Unlike fairness, care and honesty, the value of respect centres on foreseeable disagreement. If I have to respect what somebody does in a country that is not my own, even though I disagree heavily for moral reasons, does this mean there are no globally shared values? If this were the case, it would mean that no global moral framework is available to ground the 23 articles of the GCC.

In philosophy, this conundrum is called the doctrine of *moral relativism*. Such relativism has been divided into three related forms with different emphases: descriptive relativism, metaethical relativism and normative relativism.

Descriptive relativism	Descriptive relativism is a sociological or anthropological, rather than philosophical, doctrine, which is based upon observations of disparate cultures. It holds that, as a matter of fact, moral norms show considerable variation across societies: courses of action deemed permissible or even obligatory in one society may be proscribed in another.
Metaethical[4] relativism	Metaethical relativism is a philosophical doctrine that we may be tempted to adopt if we find ourselves convinced by the claims of the descriptive relativist. It maintains that there are no universal or extra-cultural moral truths: insofar as a given moral judgement can accurately be described as true or false, this position maintains that it is true or false only relative to a given society.
Normative relativism	If we subscribe to metaethical relativism, then we can only say that a particular action is wrong in *our* society, not in other societies. Our only option is to "live and let live".

A summary of these positions and how they follow from each other is shown in Figure 4.1.

Normative relativism might seem to offer us a way to resolve the apparent tension in the four-values approach, which might be thought to dovetail neatly with the normative relativist's outlook. Since we have no business interfering in or evaluating the moral codes of those from other cultures, our interactions with them, including research interactions, need to be as "light-touch" as possible. We need to treat each other fairly and with care, respect cultural differences, not impose our own values, and ensure our interactions are honest.

The problem here, however, is that just as descriptive relativism does not entail (though it may lend support to) metaethical relativism, metaethical relativism also does not entail normative relativism (and provides no support for it at all). This point is well argued by Bernard Williams (1972: 34) in his 1972 book *Morality*, in

---

[4] Metaethics is a branch of analytical philosophy which deals with higher-level questions of morality. Rather than asking how to lead a moral life, which values are appropriate to govern it, etc., metaethics asks whether such questions can be answered in the first place.

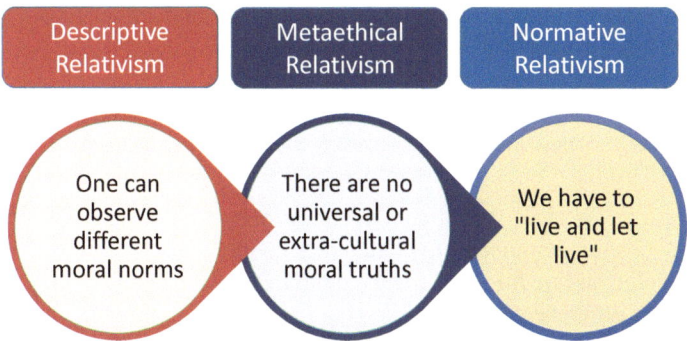

Fig. 4.1 Different types of relativism and their relationship

which he calls normative relativism "possibly the most absurd view to have been advanced even in moral philosophy". He writes:

> [T]he view is clearly inconsistent, since it makes a claim … about what is right and wrong in one's dealings with other societies, which uses a *nonrelative* sense of "right" not allowed for in [metaethical relativism].

In other words, metaethical relativism does not lead to any position, including normative relativism, which tells us what we should do in interacting with those from other cultures. This is simply because metaethical relativism itself tells us that there is no global "should". Every claim about what *we* should do has been generated and will be bound by our own culture. If the imperative to treat those from other cultures with fairness, respect, care and honesty is thought to be one that floats free of, and exists outside, any culturally bound system of values, and we want the GCC to apply globally, then we cannot be metaethical relativists.

A More Moderate Relativism

It may be that we can justify the four values by appealing to the more moderate form of relativism espoused by David Wong (2009). According to Wong, although moral norms do indeed vary across cultures, as the descriptive relativist assumes, and although there is no one single true morality, there is something that is universal about morality, and common to all particular moralities worthy of the name.

What is common is the central aim or *purpose* of morality. This aim is not itself a value in any given moral system, but rather what determines whether any given value is fit to figure in such a system. A consequence of this, and one which renders Wong's relativism more palatable for many than more extreme versions, is that although there is no single true morality, some moral systems are better than others.

Why? Because moral systems regulate interpersonal and intrapersonal conflicts. More precisely (Wong 2009: xii):

> Morality … comprises an idealized set of norms in imperatival form ("A is to do X under conditions C") abstracted from the practices and institutions of a society that serves to regulate conflicts of interest, both between persons and within the psychological economy of a single person.

For instance, Western liberal democracies stress individual rights, while other systems involve a commitment to community goods, such as those to be found in Chinese, Indian and traditional African communities (Wong 1991: 445). When we consider such differences, what reason could we have for pronouncing one culture right and the others wrong? A relativistic approach will reply, "None." Wong tells us (1991: 446):

> The argument for a relativistic answer may start with the claim that each type focuses on a good that may reasonably occupy the centre of an ethical ideal for human life. On the one hand, there is the good of belonging to and contributing to a community; on the other, there is the good of respect for the individual apart from any potential contribution to community. It would be surprising, the argument goes, if there were just one justifiable way of setting a priority with respect to the two goods. It should not be surprising, after all, if the range of human goods is simply too rich and diverse to be reconciled in just a single moral ideal.

Wong's approach may be more acceptable than an extreme, uncompromising metaethical relativism (and potentially more serviceable as a means of grounding the four GCC values), but it is not without its problems. For example, Michael Huemer (2005) points out a dilemma for any such relativism. That dilemma is revealed when we ask whether the regulation of interpersonal and intrapersonal conflicts is itself something that we should regard as good. If it is, then the regulation of such conflicts represents a value that transcends cultures, grounding any acceptable morality in any society and/or time. Hence, we would have at least one universal value rather than a form of relativism.

As a result, Wong's approach still does not give us any definite universal values (at a minimum, the four values that feature in the GCC).

Grounding the Global Applicability Thesis of the GCC in a Common Morality

In their celebrated book *Principles of Biomedical Ethics*, Tom Beauchamp and James Childress have, over the course of 34 years and seven editions, maintained that there are four principles that are particularly applicable to problems in biomedical ethics: respect for autonomy, beneficence, non-maleficence and justice. As explained above, their claim is that these principles are globally applicable because they are part of what Beauchamp and Childress call "the common morality", which is to be understood as a set of principles subscribed to by all morally committed people, whatever their culture, and whatever the time in which they live. The principles of the common morality, then, are globally applicable, in just the way that the authors of the GCC want the four values of fairness, respect, care and honesty to be.

To the rather obvious objection that no set of values/principles seems to possess the universality they ascribe to the common morality, given the observations of descriptive relativism, Beauchamp and Childress have two responses.

1. It is not the case that every principle that exists finds a home in the common morality: some principles are *purely* local.
2. More importantly, the principles of the common morality may be variously specified in different cultures.

The great benefit of the common-morality theory is that it allows an optimum balance of universality on the one hand, and variation across cultural settings on the other. There is a set of high-level values/principles that are internal to morality, but these are expressed in differing ways in particular moralities associated with particular communities.

If the supposed common morality is a set of general, unspecified values to which all who are morally committed subscribe, how do we know which these values are? Would we not first have to identify some morally committed people, and then carry out an empirical investigation into the values they hold? The most general values shared by them all would then be those that constitute the common morality. But problems loom here; circularity threatens.

How do we determine who is morally committed in the first place? Presumably, we do so by examining what values they hold. If they adhere to $value_1$, $value_2$, $value_3$, and so on up to $value_n$, then, we might want to say, they are morally committed. But this is an unacceptable, question-begging way of proceeding. If we want to find out what values the morally committed hold, and thereby find what values constitute the common morality, it is no good *defining* the morally committed in terms of their subscription to a certain definite list of values settled in advance. There either has to be an independent way of identifying the morally committed (without reference to the precise values they hold), or an independent way of determining the values of the common morality (without reference to the idea of the morally committed agent).

This last criticism was advanced by one of the authors of this book (Herissone-Kelly 2003). It was suggested that if the principles/values of the common morality are those that are internal to the concept of morality, an independent way of determining what they are will be available. What Herissone-Kelly had in mind was the sort of picture painted by Philippa Foot (2002: 7):

> [T]here are ... starting points fixed by the concept of morality. We might call them "definitional criteria" of moral good and evil, so long as it is clear that they belong to the concept of morality – to *the* definition and not to some definition which a [wo]man can choose for [her/]himself. What we say about such definitional criteria will be objectively true or false.

The thought, then, is that we perhaps can, at least in theory through a no doubt arduous and protracted process of conceptual analysis, find out what values are essential to morality. Having done that, we will be in a position to pick out those who are morally committed: they will be that group of people who subscribe to the common morality's values. But what those values are is a claim that will have been fixed quite independently of their association with the morally committed.

In the 6th edition of *Principles of Biomedical Ethics*, Beauchamp and Childress (2009: 395) take up Herissone-Kelly's suggestion when they refer to

> a plausible hypothesis, that the concept of morality contains normativity not only in the sense that morality necessarily contains *some* action guiding norms, but also in the sense of necessarily containing *specific* moral norms. These norms are privileged norms that are constitutive of morality itself .... As we have occasionally said of the four principles that provide the framework norms in this book, they are very general starting points that are fixed by morality. One way of understanding this claim is that these anchoring norms belong conceptually to morality.

Beauchamp and Childress appear to agree that an exhaustive analysis of the concept of morality is a task that lies beyond the scope of their (or any) book. However, in the absence of such an analysis they make two ingenious suggestions about how we might determine the constituents of the common morality and so establish the global applicability thesis. These suggestions, if they are acceptable, provide a legitimate way of using the identification of the morally committed in order to establish the constituents of the common morality.

The first suggestion is that some values are *internal* to the concept of morality and thus globally applicable. As a result, they can be legitimately employed in the sort of cross-cultural context that the GCC is designed to cover.

The second suggestion is that once it is agreed that there will be certain (as yet unidentified) values that are constitutive of morality, one can with confidence identify at least *some* such privileged values, even in the absence of anything like a *comprehensive,* conceptual analysis.

Beauchamp and Childress maintain that one principle that unequivocally appears internal to the concept of morality is that of non-maleficence, or do no harm.[5] This is an enormously plausible view; it seems unthinkable that any system lacking this principle could even be counted as a candidate for a morality.

Armed with the knowledge that non-maleficence is a foundational principle, and so a principle to which the morally committed will pay heed, Beauchamp and Childress (2009) go on to suggest that we can first identify a large sample of agents who adhere to that norm, and then determine what other general, non-specified principles they all hold in common. Those further general principles will be the remaining constituents of the common morality.

In this scenario, partial analysis of the concept of morality reveals that one of its internal principles is non-maleficence. Morally committed persons can then be identified by their subscription to this principle (and not by their adhering to some principle that we just happen to like to think of as very important).

The next stage in this method would involve determining what *other* principles all those morally committed persons champion. In this way, one could discern the

---

[5] "Do no harm" is phrased as a principle, not as a value. Care is a value that includes "do no harm".

constituents of the common morality without undertaking the rather forbidding chore of a full-blown, exhaustive analysis of the concept of morality itself.

Beyond this thought experiment, however, we would still have to account for what appear to be counter-examples to Beauchamp and Childress's claim that their four principles are globally applicable. Such counter-examples have been produced, especially against the principle of respect for autonomy, which is regarded as bound to Western cultures (Huxtable 2013; Kara 2007, Cheng-Tek Tai 2013, Kiak Min 2017). We want to examine one such counter-example.

R.E. Florida (1996), in an article entitled "Buddhism and the Four Principles", insists that no principle of respect for autonomy is to be found in Buddhist cultures at all, due to Buddhism's metaphysic of "co-conditioned causality". The thought seems to be that autonomy is not going to show up as a value, and so as something especially worthy of respect, in a culture that holds to a conceptual scheme in which there is no genuine separation between what those outside that culture call "individuals" (any apparent separation being at best an illusion).

The interesting thing about the Buddhist example is that it would be very difficult to argue that Buddhist culture is not fundamentally committed to non-maleficence. The notion looms exceptionally large in Buddhist ethics, in the shape of the norm of *ahimsa*, or non-harm. Those who are committed to Buddhist ethics, then, are, by Beauchamp and Childress's vision, morally committed. And yet, if Florida is right, subscription to the principle of non-maleficence is not universally accompanied by subscription to the principle of respect for autonomy. Hence, at least one element of the group of four principles chosen by Beauchamp and Childress, namely respect for autonomy, does not seem to belong to the common morality.

Conclusion

The most promising approach based on values/principles to claim global applicability for its framework has a hole. As the example of Buddhist ethics shows, respect for individual autonomy cannot be regarded as a globally applicable value/principle that all morally committed people would subscribe to. Respect for autonomy, with the focus on *individual* autonomy, as understood by Beauchamp and Childress (2013: Chapter 4) does not seem to qualify for global applicability, as the Buddhist counter-example is solid. But that does not mean other values systems must fail.

If key values are, as argued above, internal to morality, the four-values approach developed for the GCC is valid until falsified with rigorous counter-examples, such as the Florida example against respect for autonomy.

References

Beauchamp TL, Childress JF (2009) Principles of biomedical ethics, 6th edn. Oxford University Press, New York

Beauchamp TL, Childress JF (2013) Principles of biomedical ethics, 7th edn. Oxford University Press, New York

Cheng-Tek Tai M (2013) Western or Eastern principles in globalized bioethics? An Asian perspective view. Tzu Chi Medical Journal 25(1):64–67

Florida RE (1996) Buddhism and the four principles. In: Gillon R (ed) Principles of healthcare ethics. John Wiley and Sons, Chichester, p 105–116

Foot P (2002) Moral dilemmas. Oxford University Press, Oxford

Herissone-Kelly P (2003) The principlist approach to bioethics, and its stormy journey overseas. In: Häyry M, Takala T (eds) Scratching the surface of bioethics. Rodopi, Amsterdam and New York, p 65–77

Huemer M (2005) Ethical intuitionism. Palgrave Macmillan, New York

Huxtable R (2013) For and against the four principles of biomedical ethics. Clinical Ethics 8(2–3):39–43

Kara M A (2007) Applicability of the principle of respect for autonomy: the perspective of Turkey. Journal of Medical Ethics 33(11):627–630

Kiak Min MT (2017) Beyond a Western bioethics in Asia and its implication on autonomy. The New Bioethics 23(2):154–164

Williams B (1972) Morality: an introduction to ethics. Cambridge University Press, Cambridge

Wong D (1991) Relativism. In: Singer P (ed) A companion to ethics. Blackwell, Oxford, p 442–450

Wong D (2009) Natural moralities: a defense of pluralistic relativism. Oxford University Press, New York

Chapter 5
Exploitation Risks in Collaborative International Research

Abstract Ethics dumping occurs in collaborative international research when people, communities, animals and/or environments are exploited by researchers. Exploitation is made possible by serious poverty and extreme power differentials between researchers from high-income countries and research stakeholders from low- and middle-income countries (LMICs). To prevent its occurrence, the risks of exploitation have to be tackled. This chapter describes 88 risks identified for collaborative international research, categorized according to four values: fairness, respect, care and honesty. The risks were identified in a broad-based consultative exercise, which included more than 30 members and chairs of ethics committees in LMICs, representatives from vulnerable populations in LMICs, and an open call for case studies of exploitation. The findings of the exercise contributed to the development of the Global Code of Conduct for Research in Resource-Poor Settings.

Keywords Exploitation · Ethics dumping · Collaborative research · Vulnerability · Research ethics · Ethics codes

Ethics dumping[1] occurs in collaborative international research when people, communities, animals and/or environments are exploited by researchers. In order to prevent ethics dumping, such exploitation needs to stop. This chapter describes our investigation into the risks of exploitation in low- and middle-income countries (LMICs), uncovering what makes exploitation more likely to occur due to vulnerabilities that can be exploited, either knowingly or unknowingly.

This undertaking was vital for the development of a Global Code of Conduct for Research in Resource-Poor Settings (GCC) that can address real-world risks for exploitation in research. Many such risks are not well described in the literature, and hence there was an empirical component to our activities. Furthermore, this process was necessary to ensure that the GCC was more than a compilation of existing

---

[1] The export of unethical research from a high-income setting to a resource-poor setting with weaker compliance structures or legal governance mechanisms.

codes, most of which had not been written with LMIC-HIC (high-income country) collaborations in mind.

The Nature of Exploitation

The potential to be exploited is part of the human condition. Exploiters take advantage of others' vulnerabilities to promote their own interests (Hughes 2010). While there is a morally neutral sense of exploitation (the exploitation of natural talents to create art, for example), the term is generally used to describe a moral failing. Exploitation of people is very often unjust, unfair, harmful or just plain wrong. What is it, then, that distinguishes morally unacceptable exploitation from neutral exploitation?

Some argue that exploitation is wrong because it is coercive (Schwartz 1995). If the only way for a woman in an LMIC to access antiretroviral drugs to prevent the transmission of HIV to her unborn baby is to participate in a placebo-controlled clinical trial,[2] despite the existence of a proven standard of care,[3] then one could say she has been coerced into enrolling (Annas and Grodin 1998). In this sense, exploitation occurs where one party takes advantage of another by making them an offer they cannot refuse; they are then coerced to accept simply because there is no alternative. Others argue that exploitation is wrong because it treats human beings as means rather than ends (Wood 1995). In other words, exploitation instrumentalizes people. Yet others claim that exploitation is wrong because it disadvantages the vulnerable (Macklin 2003).

Our investigation was concerned with the risks or vulnerabilities for exploitation, so we adopted Macklin's definition of exploitation. However, it is important to bear in mind that situations that are conducive to exploitation do not necessarily lead to exploitation. For instance, if a pharmaceutical company is due to test new antiretroviral drugs to prevent the transmission of HIV to unborn babies, and the company operates in a country where poor mothers have no or very limited access to health care, it does not mean that exploitation will necessarily occur. The company may decide not to exploit vulnerable research participants and offer the accepted standard of care to those in the control arm, rather than a placebo.

Exploitation usually requires a moral decision on the part of the potential exploiter, but it can also occur through ignorance. Whether intended or unintended, the effects of exploitation are the same for the exploited. Hence, ignorance is not a legitimate justification for exploitation. Uncovering the primary risks of exploitation

---

[2] A placebo-controlled trial involves some participants being given a medicine with active ingredients, for instance a new drug against malaria, while others, the control group, are given a substance that should have no effect (the placebo), so that the outcomes can be compared.

[3] One speaks of a proven standard of care when a treatment already exists for the illness under consideration in a trial. Hence, the ethical demand of testing any new drug against an existing one rather than a placebo is known as the "standard of care" debate.

can help to increase awareness but, in our case, it also ensures that the GCC is designed in such a way that researchers are compelled to consider these factors. It is a unique facet of the GCC that it focuses the attention of researchers directly upon the primary risks of exploitation in collaborative HIC-LMIC research. This could only be achieved via thorough exploration of the risks from many perspectives, both top-down and bottom-up.

Our Method

The aim of this investigation was to identify the critical vulnerabilities that engender susceptibility to exploitation in LMIC-HIC collaborative research. Investigation of this vast subject would be impossible from a traditional literature-based approach, or through investigation in a single geographical region. Many of these vulnerabilities are poorly represented in the literature, and they can differ between countries, cultures and types of research. For example, clinical trials, social science, animal experiments, environmental science and research in emergency settings may pose a diverse array of risks that are largely determined by the local context. Consequently, a creative approach to data collection was needed to capture as many risks and vulnerabilities as possible.

In this regard it was very helpful that the interdisciplinary TRUST project consortium comprised multilevel ethics bodies, policy advisers and policymakers, civil society organizations, funding organizations, industry and academic scholars from a range of disciplines. With input from each of these perspectives, a broad-based consultative exercise[4] was possible which included input from these collaborators as well as more than 30 members and chairs of ethics committees in LMICs, representatives from vulnerable populations in LMICs, and an open call for case studies of exploitation in research in LMICs (Chapter 6).

For example, extensive input from members and chairs of ethics committees was sought in both India and Kenya. In India, the Forum for Ethics Review Committees in India (FERCI) hosted a two-day workshop in Mumbai on 11 and 12 March 2016. At this workshop, approximately 30 leading bioethicists from around India came together to share their experiences and discuss cases of exploitation in research. In Nairobi on 23 and 24 May 2016, three esteemed chairs of national ethics committees shared their experiences and opinions about the primary ethical challenges for LMIC-HIC collaborative research in Kenya. Findings from both events revealed multiple risks of exploitation that are characteristic of research in some LMIC settings. These included traditional requirements for appropriate community

---

[4] This type of consultative exercise is of proven value in the development of ethical codes that are broadly representative and can have wide-ranging impact. For example, the principles of the "Three Rs", which are globally accepted as a reasonable measure for ethical conduct in animal research, arose from a broad consultation with stakeholders undertaken by Russell and Burch in the 1950s. See Russell et al. (1959).

consultations and permissions, and specific cultural beliefs and customs that must be respected.

Ongoing consultation with representatives from two vulnerable groups that have first-hand accounts of the risks for exploitation were undertaken. From Nairobi, Kenya, sex worker peer educators and, from South Africa, members of the San community shared their experiences of being the subjects of exploitation and their opinions about how they want to be treated in future. Among many other insights, both groups described a lack of benefits from research projects (which are often highly beneficial to the researchers), as well as risks of stigmatization from the manner in which they were involved in the study.

12 months of in-depth and far-reaching investigation produced a considerable amount of data (Chapter 6). From this data, individual vulnerabilities and risks of exploitation were extracted, organized and tabulated on an Excel spreadsheet with source details and descriptions of the vulnerability or risk. Care was taken to ensure that each individual entry was based upon real-world experience rather than hypothetical suppositions. Our lists were compared with risks mentioned in the literature and, where necessary, additional information sought to address gaps.

Once collated, the raw data was streamlined to group similar vulnerabilities together. For instance, there were many different examples of how people living in resource-poor circumstances may be unfairly enticed to participate in research by the prospect of payment or reward. Such examples were grouped under the label "undue inducement". Further thematic analysis resulted in distinctions between the various potential subjects of exploitation, or levels of risk for exploitation (persons, institutions,[5] local communities, countries, animals and the environment). In the final stage of the analysis the vulnerabilities were grouped according to the four values of fairness, respect, care and honesty.

Our Findings

The remainder of this chapter is devoted to presenting and explaining our findings. For each value, an exploitation risk table details the main risks for persons, institutions, local communities, countries, animals and the environment. Each entry on the tables describes a vulnerability that could lead to exploitation (deliberate or unintentional) in LMIC-HIC research collaborations and all are grounded in real-world experience. Additionally, for each value, certain examples are described in more detail to further illustrate the risks.

It has to be noted that some entries could have been linked to more than one value. For instance, if a research participant suffered from a therapeutic misconception, the researcher might not have taken enough care to explain that research is different from treatment because s/he was not aware that this might be problematic in some settings, or otherwise because s/he deliberately and dishonestly wanted to avoid explaining the difference, in which case the value of honesty would have been violated. To avoid overburdening the tables, we made a decision to prioritize one value in each case.

---

[5] "Institutions" includes local researchers as well as their organizations.

Fairness

Our data revealed many risks for exploitation that might be categorized as issues of fairness (or unfairness) that are varied in nature and pertain to different aspects of fairness. Philosophers commonly distinguish between different types of justice or fairness (Pogge 2006) (Chapter 3), but the most relevant fairness concepts for global research ethics are *fairness in exchange* and *corrective fairness*.

Fairness in exchange concerns the equity of transactions that occur between parties. In collaborative research, ventures should aim to be mutually beneficial. Where the collaboration is between HIC and LMIC partners, typical fairness in exchange issues might include:

The relevance of the research to local needs
Whether reasonable benefit sharing is taking place
Whether LMIC researchers are involved in meaningful ways

Table 5.1 shows the primary risks related to fairness in exchange for persons, institutions, communities, countries and the environment.

Table 5.1 Primary risks for fairness in exchange

Level of risk	Nature of risk
Personal	• In medical research: - Multiple trial enrolment - No post-study access to treatment • In all research: - Undue inducement - No access to results or benefits of research
Researcher/ institutional	• Research priorities driven by HIC partners: - Mismatch to local research needs • Poor representation of LMIC (host) partners on research teams: - Responsible for menial tasks only - Not acknowledged or represented appropriately in publications • "Helicopter research" by HIC partners: - No knowledge transfer or capacity building/strengthening
Community	• Research priorities driven by HIC partners: - Mismatch to local research needs • Little or no input from marginalized communities into research • Undue inducement • No benefit sharing or feedback • Support for foreign-sponsored research drains local system of staff
Country	• No universal access to health care for population: - Differences in standards of "usual" care • Placebo-controlled trials approved • Support for foreign-sponsored research drains local systems and resources • Medical science research shaped by the "para state"
Animal	
Environmental	• Study leads to reduction of natural resources • Lack of benefit sharing for the environment

A primary risk in the exploitation of individuals and communities is that they may not have access to the results or benefits of research. This occurs when the research is designed to benefit people in other countries or settings and the individuals who contributed to the study never get a chance to benefit from it. This happened with a clinical study of the hepatitis B vaccine in Kenya. Although the research was undertaken in Kenya, for many years afterwards people in Kenya could not afford to purchase the vaccine and therefore could not benefit (Bhatt 2016). When research aims are driven by, and in the interests of, high-income researchers or institutions with no real benefit to the local community or participants, we must ask why it is being conducted there.

Local LMIC researchers are exploited when used only for tasks such as data collection, or when, having participated in a research project, they are then not properly represented, or not represented at all, in subsequent publications. Local environments are exploited when environmental studies fail to benefit them. Research agreements focused on the use of biodiversity and traditional knowledge typically ignore the environmental component, and the common approaches to benefit sharing from research activities include only humans (Stone 2010).

Corrective fairness presupposes the availability of legal instruments and access to mechanisms for righting wrongs (e.g. a complaints procedure, a court or an ethics committee). For instance, if no research ethics structure exists in the host country, corrective fairness is limited to the research ethics structure in the HIC country, which may not have the capacity to make culturally sensitive decisions. Table 5.2

Table 5.2 Primary risks for corrective fairness

Level of risk	Nature of risk
Personal	• Difficult or no access to legal system or legal aid • Human rights violations not taken up by civil society
Researcher/ institutional	• Lack of protection of IPR for LMIC institutions • Lack of clear standards for operating systems and timelines for RECs • No capacity/procedures for study oversight to ensure compliance with REC decisions
Community	• Lack of protection of IPR or traditional knowledge for local communities • Human rights violations not taken up by civil society • Absence of systems for community approvals
Country	• No relevant legal instruments for ethics committees • Poor research governance frameworks to ensure adherence to ethical standards • No cross-border legal recourse in cases of exploitation • Discriminatory laws that may create stigmatized minorities
Animal	• Variations in regulatory standards for animal experimentation • Inadequate systems to ensure compliance with animal welfare standards
Environmental	• Variations in governance of natural resources • Variations in procedural rights • Environmental protection not well policed by civil society

shows the primary risks related to corrective fairness for persons, institutions, communities, countries, animals and the environment.

Individuals who are harmed by their participation in research may have no means of seeking retribution or compensation if they cannot afford legal representation and there is no form of legal aid. For communities, a lack of awareness and expertise, or too much trust in the HIC researchers, may lead to the loss of intellectual property rights (IPRs) to local knowledge and resources. At a national and international level, researchers from HICs who choose to ignore or flout the research ethics and legal requirements in the host LMIC can be difficult to police. This is especially problematic in localities where there is a lack of resources and/or infrastructure to ensure ethical compliance through the entire research process and where the home institutions in HICs do not ensure that their employees comply with requirements.

Respect

Respect requires an acceptance of customs and cultures that may be different from one's own, and a commitment not to behave in a way that causes offence. One may need to abide by decisions or ways of approaching matters with which one disagrees. This can be problematic, especially if local customs are illegal or perceived as dangerous.[6] However, respect is important in LMIC-HIC collaborations, and there are many possible ways of showing respect that do not create conflicts of conscience. For instance, HIC researchers should not insist that LMIC ethics committees accept the ethics approval submission in the HIC's preferred format, but should rather conform with the format preferred by the LMIC committee. Table 5.3 shows the primary risks related to respect for persons, institutions, communities, countries, animals and the environment.

Local LMIC customs, traditions, and religious and spiritual beliefs may be very different from those of the HIC researcher. For example, from an African cultural point of view, human body parts are sacred, whether they are obtained from living or deceased persons. Hence, the removal of blood or other body parts for research may have a profound impact that needs to be acknowledged and addressed in a manner that is sensitive to the wishes of the local community. A liberal interpretation of autonomy, i.e. *individual* autonomy, prevails in HICs but may not be easily transferred to LMIC settings where "community" or "group autonomy" is also highly valued. Furthermore, in some settings it might be deemed rude for a research participant to say "no" or to ask questions about the research. In other situations, people may be too afraid or unconfident to do so. Either way, the power imbalance between researcher and research participant can impact upon the consent process.

---

[6] For instance, if a researcher learns that female genital mutilation is being used as a "cure" for diarrhoea in female babies, respecting this approach to health care is likely to be the wrong decision, particularly as the practice is likely illegal. At the very least such a decision would leave the researcher with a serious conflict of conscience (Luc and Altare 2018).

Table 5.3 Primary risks for respect

Level of risk	Nature of risk
Personal	• Unequal power relations • Tendency to defer to authorities • Individual spiritual and religious priorities incompatible with or ignored by HIC partners • Researchers and/or ethics committees deciding "what is best"
Researcher/institutional	• Research protocol and papers imported from HIC partners and not tailored to local needs • Ethical approval sought only from HIC partner

Community	• Diverse interpretations of important values
	• Local requirements for effective community engagement ignored
	• Diverse ethical priorities for matters such as:
	  - gender equality
	  - sexual relations
	• Particular spiritual and religious priorities incompatible with or ignored by Northern partners
	• Localized social effects from research team presence
	• Local customs that may violate laws of the country and/or human rights

Country	• Research protocols and practices which fail to take account of national traditions and legislation
Animal	• Variations in customs, norms and attitudes regarding animal welfare and inhumane practices
Environmental	• Variations in customs, norms and attitudes regarding the environment

Animals and environments are also at risk of exploitation because of variations in customs and norms. What is considered "animal cruelty" or "inhumane practice" in animal experimentation varies greatly between cultures. Additionally, some animals are awarded greater protection in certain cultures than others, for example, dogs and cats in the United Kingdom and cows in India. Animal experimentation on non-human primates is particularly controversial in most countries, but in some certain non-human primates are viewed as "pests" (Hill and Webber 2010). Different partners in collaborative research may have different philosophies related to the environment. Environmental protection is sometimes regarded as a colonial construct that has negative impacts on local communities in LMICs, and research agendas likewise. There may therefore be a philosophical or paradigmatic difference between research partners that needs to be identified and addressed.

Care

Researchers who take good care in their research combine two elements: they care about research participants, in the sense that they are important to them, and they feel responsible for the welfare of those who contribute to their research, or might suffer as a result of it. In work with vulnerable communities, this might, for example, entail the tailoring of informed consent procedures to local requirements (language, literacy, education levels) to achieve genuine understanding. Table 5.4 shows the primary risks related to care for persons, institutions, communities, countries, animals and the environment.

At the individual level, variations in spoken language, understanding, levels of literacy and use of terminology are just some of the issues that can lead to exploitation. The number of different ways in which individuals can suffer harm as a result of their involvement in research is vast. At the community level, the mere presence of a research team can have a great impact upon a local community. Research teams require food and accommodation, purchase local goods and services, and form relationships with local people.

At a national and international level, the rapid emergence of high-risk applications of technologies such as genome editing[7] challenges not only safety risk assessments but also existing governance tools. This creates an environment where risky experiments might be carried out in countries with an inadequate legal framework,

Table 5.4 Primary risks for care

Level of risk	Nature of risk
Personal	• In medical research: therapeutic misconception • Misunderstanding of research aims • Procedures for informed consent not tailored to individual • Lack of possible actions to address adverse effects of participation • Direct risks, such as physical side effects • Indirect risks, such as stigmatization
Researcher/institutional	• No host country research ethics structures or inappropriate match with requirements • No capacity in existing REC • REC members are poorly trained and lack specialized expertise to review ALL types of research protocols • REC meetings are either too few or too sporadic • REC does not have local or national government or ministry support to conduct its activities
Community	• Localized physical effects from research team presence
Country	• Insufficient data security measures • Insufficient safeguarding protocols • Lack of risk management approaches to biosafety • Lack of risk management approaches to biosecurity
Animal	• Animal research centres established in countries where regulation is less stringent • Lack of resources for humane animal care
Environmental	• Inadequate consideration of unintended consequences for biodiversity and the environment • Inadequate consideration of local environmental contexts • Disregard for long-term effects upon local environment • Lack of resources for environmental protection • Insufficient information for assessment of environmental effects

---

[7] For example, applying genome editing technologies to human embryonic stem cells.

or in countries where although legal frameworks exist, their effective implementation is prevented by limited resources.

Most LMICs have processes in place for ethical approval of research, but they are hindered by resource issues. Even where research ethics committees are well established, they may have limited capacity and expertise. For effective governance, the wide-ranging and dynamic nature of research requires extensive and cutting-edge expertise from research ethics committees. This can be particularly difficult in resource-poor areas.

Inadequate environmental information in LMICs means that research decisions and directions may be developed in a vacuum and result in long-term harm. For example, research programmes may introduce exotic species that deplete water resources, displace traditional varieties − thereby impacting upon agricultural biodiversity or "escape" and become invasive, thus threatening biodiversity.

Honesty

In all cultures and nations, "do not lie" is a basic rule of ethical human interaction. However, the value of honesty has a broader scope in the context of global research ethics. Lying is only one possible contravention. In research ethics it is equally unacceptable, for instance, to omit important information from an informed consent process: that is, to fail to be transparent. The duties of honesty also include, most prominently, research integrity, which entails practices such as giving due credit for contributions and refraining from the manipulation of data and the misappropriation of research funds. In this section, we distinguish between these two forms of honesty and divide the risks into those that are mainly related to transparency and those that are more readily aligned with integrity. Table 5.5 sets out the primary risks to

Table 5.5 Primary risks for honesty through transparency

Level of risk	Nature of risk

Personal	• Inability of participant to provide fully informed individual consent:
	    - Incomplete information provided
	    - Information provided in an inappropriate format
	• Potential effects of participation not fully explained
	• Dual roles of researcher

Researcher/ institutional	• REC not fully independent • Cryptic research procedures

Community	Inability to provide fully informed community consent:
	    - Incomplete information provided
	    - Information provided in inappropriate format
	• Potential effects of participation not fully explained
	• Dual roles of researcher

Country	• Lack of data sharing
Animal	
Environmental	• Incomplete information about potential risks or harm to the environment

honesty through transparency for persons, institutions, communities, countries and the environment.

There are many ways in which the informed consent process can be inadequate. For example, where there are omissions and/or inappropriate or misleading language for the context in which consent is being sought; when potential participants (some of whom may be illiterate) are not taken through any kind of suitable consent process but are, instead, provided with written information sheets to take home and told to come back with a signed consent form; or when information does not fully explain the potential (possibly harmful) consequences of participation. Misunderstanding can lead to a violation of trust: for example, where the researchers are also aid workers or health care providers, potential participants may believe that they have to participate in order to receive the aid or treatment.

A lack of transparency concerning research processes can make it very difficult to hold anyone accountable when things go wrong. For example, where numerous bodies are engaged in collaborative research and their separate activities and responsibilities are not clear, then it may be impossible to say where things have gone wrong and who is responsible.

Honesty is also the foundation of research integrity. Honesty is essential in all aspects of research, including

> in the presentation of research goals, intentions and findings; in reporting on research methods and procedures; in gathering data; in using and acknowledging the work of other researchers; and in conveying valid interpretations and making justifiable claims based on research findings (Universities UK 2015).

Table 5.6 shows the primary risks for honesty through integrity for persons, institutions, communities, countries, animals and the environment.

Research misconduct can happen anywhere; it is not unique to collaborative ventures in LMICs. However, variations in customs and a lack of facilities to ensure oversight of compliance with research ethics might encourage unscrupulous behaviour. In our investigation we found that people can be put at risk if sensitive personal data is not sufficiently protected or exploited; when, for example, blood samples or

Table 5.6 Primary risks for honesty through integrity

Level of risk	Nature of risk
Personal	• Personal data protection breaches • Unauthorized secondary use of samples • Use of samples for commercial purposes without consent • Deliberate withholding of information • Deliberate obfuscation of research aims
Researcher/ institutional	• Bribery in existing REC • Ingrained institutional unethical practices or institutional culture of disregard for legal requirements
Community	• Dishonoured commitments
Country	• Data sharing without consent because of lack of strict privacy arrangements
Animal	• Deliberate obfuscation of experimental conditions
Environmental	• Results from HIC research inappropriately applied in LMIC context

data are sold for profit without the knowledge or consent of the participants. In some institutions, certain unethical practices may become the norm, and even those workers who object may feel pressurized to comply. Additionally, researchers may break promises routinely; where, for example, the researchers have promised to return with feedback about the research and then fail to do so. A lack of integrity can also threaten animal welfare. Where standards of animal housing and care are not high, obfuscation about the conditions in which the animals are kept may occur for two main reasons: first, to ease the process of gaining ethical approval from the HIC partner, and secondly, so as not to jeopardize publication of the experimental results.[8]

Conclusion

Our findings revealed 88 separate risks for exploitation but in this conclusion, we draw attention to three factors that underpin many of the individual risks, namely:

Extreme differentials in available income, i.e. serious poverty
Extreme differentials in power
Past history of colonialism

Serious Poverty

The most obvious risk of exploitation in LMIC-HIC collaborative research is the extreme divergence in levels of affluence across the world. Many of the individual points in the risk table have their origins in extreme poverty. Researchers cannot solve this problem, but they can show heightened awareness of it and try their best to promote local improvements, for example by equitably involving local researchers, by focusing their research on local research needs and by obtaining input from local populations.

Extreme Differentials in Power

The relationship between wealth and power, and the differentials in power between those who are wealthy and those who live in poverty, is a topic that has filled books, halls and television programmes. However, other factors also play a part in

---

[8] Many academic journals that publish results from animal experimentation stipulate requirements that the studies have been conducted in a manner that is consistent with high ethical standards such as EU Directive 2010/63 (EU 2010).

maintaining power differentials. To give one striking example, the United Nations Security Council has 15 members. Of these, ten are elected by the General Assembly for periods of two years while five (China, France, Russia, the United Kingdom and the United States) have been permanent members with "veto power" since 1946. More than 60 United Nations member states have never been members of the Security Council and hence have never even had voting rights, let alone veto power.

Past History of Colonialism

Does the history of colonialism still bear upon research today? We believe it does, as can be seen from the experience of indigenous peoples in research. Linda Tuhiwai Smith has powerfully shown that indigenous peoples often consider research a "dirty word". She describes how "imperialism frames the indigenous experience" and how "indigenous peoples had to challenge, understand and have a shared language for talking about ... colonialism" (Smith 2012).

We close this chapter with a comment from the TRUST gender adviser, Prof. Fatima Alvarez-Castillo (2016):

> A culture's worldview, expressed in language, contains norms and values about power and relations of power. For example, the word "expert" imbues persons with authority and assigns higher credibility to their claims than those of non-experts. The public is expected to defer to their opinions on matters of their expertise. It was not until about the 1960s when the usual understanding of expertise was challenged by feminists, who argued that unschooled women have more expertise about their own situation than the experts. This ushered in a new research philosophy that valorizes poor women's stories and their own versions of their realities.

References

Alvarez-Castillo F (2016) Gender sensitivity: writing and language. In: Chatfield K, Schroeder D, Kimani J (eds) Nairobi plenary meeting report, TRUST Project. http://trust-project.eu/wp-content/uploads/2016/11/Meeting-Report-TRUST-Nairobi-Final.pdf

Annas G, Grodin M (1998) Human rights and maternal-fetal HIV transmission prevention trials in Africa. American Journal of Public Health 88(4):560–563

Bhatt K (2016) Concerns for Kenyan National Bioethics Committee when approving North-South collaborative projects. In: Chatfield K, Schroeder D, Kimani J (eds) Nairobi plenary meeting report, TRUST Project. http://trust-project.eu/wp-content/uploads/2016/11/Meeting-Report-TRUST-Nairobi-Final.pdf

EU (2010) Directive 2010/63/EU of the European Parliament and of the Council of 22 September 2010 on the protection of animals used for scientific purposes (text with EEA relevance). OJ L 276/33. https://eur-lex.europa.eu/legal-content/EN/TXT/?uri=celex%3A32010L0063

Hill CM, Webber AD (2010) Perceptions of nonhuman primates in human–wildlife conflict scenarios. American Journal of Primatology 72(10):919–924

Hughes J (2010) European textbook on ethics in research. European Commission, Brussels

Luc G, Altare C. (2018) Social science research in a humanitarian emergency context. In: Schroeder D, Cook J, Hirsch F, Fenet S, Muthuswamy V (eds) Ethics dumping: case studies from North-South research collaborations, Springer Briefs in Research and Innovation Governance, Berlin, p 9-14

Macklin R (2003) Vulnerability and protection. Bioethics 17(5–6):472–486

Pogge T (2006) Justice. In: Borchert DM (ed) Encyclopedia of philosophy, 2nd edn, vol 4. Macmillan Reference, Detroit, pp 862–870

Russell WMS, Burch RL, Hume CW (1959) The principles of humane experimental technique. Methuen & Co, London

Schwartz J (1995) What's wrong with exploitation? Nous 29:158–164

Stone CD (2010) Should trees have standing? Law, morality, and the environment, 3rd edn. Oxford University Press, Oxford

Smith, LT (2012) Decolonizing methodologies: research and indigenous peoples. Zed Books, London

Universities UK (2015) The concordat to support research integrity. Universities UK, London

Wood A (1995) Exploitation. Social Philosophy and Policy 12:150–151

Chapter 6
How the Global Code of Conduct Was Built

Abstract How can an ethics code achieve impact? The answer is twofold. First, through adoption by influential research funders, who then make it mandatory for their award recipients. This is the case with the Global Code of Conduct for Research in Resource-Poor Settings, which was adopted by both the European Commission and the European and Developing Countries Clinical Trials Partnership shortly after its launch in 2018. Second, an ethics code can achieve impact when researchers use it for guidance whether it is compulsory or not. This is most likely to happen with codes that were developed transparently with all research stakeholders involved. This chapter will outline how the GCC was developed, and in particular how external stakeholders were systematically engaged, how existing codes were carefully analysed and built upon, and who the early adopters were.

Keywords Research ethics · Ethics dumping · Stakeholder engagement · Ethics codes

Introduction

The earliest research ethics codes were written solely for researchers:

> The Nuremberg Code (1949) and the original Declaration of Helsinki (1964) made no mention of committee review; these documents placed on the investigator all responsibility for safeguarding the rights and welfare of research subjects. (Levine 2004: 2312)

In 1966, the surgeon general of the US Public Health Service issued a policy statement requesting the establishment of research ethics review committees or institutional review boards (Levine 2004: 2312). At this point in history, ethics codes would have had two main target audiences: researchers and research ethics committees.

The Global Code of Conduct for Research in Resource-Poor Settings (GCC) was developed from the start with three audiences in mind: researchers, research ethics

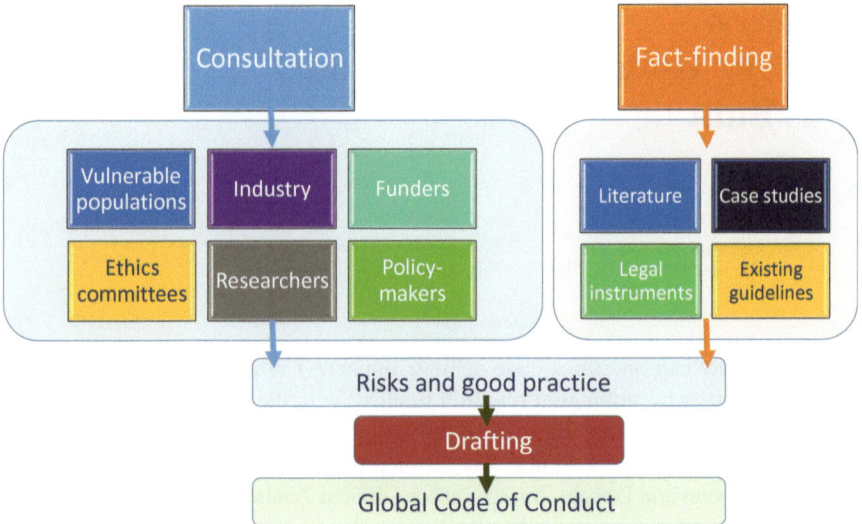

Fig. 6.1 Input into the GCC

committees and research participants (and/or their communities and support groups). To achieve a good match with these anticipated audiences, it was essential that all relevant stakeholders[1], in particular highly vulnerable populations, be included at all stages of the drafting process of the GCC. This inclusion was later praised by the Deputy Director-General for Research and Innovation of the European Commission (EC) as "impressive" and

> a Horizon 2020 success story, [which] demonstrates that, in order to actively combat ethics dumping,[2] a coordination of stakeholder efforts is required. Moreover, following the example of TRUST,[3] such efforts should be based on a bottom-up approach that empowers local communities involved, as equal partners. (Burtscher 2018: 1)

A major benefit of the bottom-up approach is that it resulted in a short, clear code that is focused on practical matters and accessible to nonspecialists. The development process consisted of a range of activities, which are summarized in Figure 6.1.

Case studies were collected prior to the drafting of the GCC and published in a book entitled *Ethics Dumping: Case Studies from North-South Research Collaborations* (Schroeder et al. 2018). The foundation of the GCC, the risk matrix, has been introduced in Chapter 5 of this book.

---

[1] "Stakeholders" is an increasingly contested term, as it may imply that all parties hold an equal stake. Some prefer the term "actors", yet this brings its own complexities. While acknowledging the debate, we use the well-established term "stakeholders" throughout.

[2] The export of unethical research from a high-income setting to a resource-poor setting with weaker compliance structures or legal governance mechanisms.

[3] The TRUST project (http://trust-project.eu/) was funded by the European Commission from 2015 to 2018. One of its outputs was the GCC.

This chapter will outline the following steps in the development of the GCC:

- The consultations with various stakeholders over two and a half years
- The comparative analysis of existing guidelines and relevant legal instruments
- The drafting process.

Meaningful Consultation with Diverse Stakeholders

"They could take you out for coffee and call it consultation!" (Youdelis 2016)

Consultation, engagement, community engagement, guided discussions, focus groups and interviews: these are all means of obtaining relevant input from stakeholders prior to action. When a problem affects a range of people or groups and a feasible and implementable solution is sought, stakeholder input and community engagement are essential (Hebert et al. 2009; Cook 2008; Bassler et al. 2008; Dunn 2011).

This section shows how a wide range of stakeholders were consulted on ethics dumping concerns and potential solutions, with the specific aim of drafting the GCC. (Chapter 8 provides more general advice on community engagement with vulnerable populations (Chapter 8).)

Broad Consultation

Governance mechanisms such as ethics codes require evidence of legitimacy. Why should a particular ethics code be followed by researchers? The answer that can be given for the GCC is fourfold. First, the funder that supported the development of the GCC – the European Commission – requested a *new* code to guard against ethics dumping. Hence, instead of engaging in a long-term process to negotiate the addition of specific sections on international collaborative research to existing ethics codes and governance mechanisms, a funder with an interest in the output opted for a new and independent code. Second, in 2015 the TRUST consortium's bid was chosen by peer reviewers from a range of proposals to tackle ethics dumping. The criteria for the selection were excellence and impact, as well as the quality and efficiency of the proposed implementation (Horizon 2020 nd). Third, upon the completion and launch of the GCC in 2018, the EC ethics and integrity sector and the EC legal department assessed the code and the decision was taken to make it a mandatory reference document for European Union (EU) framework programmes (Burtscher 2018). However, the most important element for the GCC's credibility may be the fourth element, the fact that the TRUST consortium made every possible effort to engage *all* relevant stakeholders across five continents in the development of the GCC.

In the effort to reach all relevant stakeholders, the TRUST consortium first had to agree upon who needed to be consulted on ethics dumping in research. The following six groups were identified:

1. The research process starts with *research policymakers*, who set the parameters for research activities. For instance, in the European Union, research aimed at human cloning for reproductive purposes is forbidden (European Commission 2013), which means it is outside the activity range of researchers.
2. A second highly influential stakeholder group consists of *research funders*. Without specific funding, most research is not possible. Whether research funding is provided by industry, charitable foundations or state-funded research programmes makes no significant difference. All funders are of particular importance in tackling ethics dumping, as they often set specific ethical rules that the researchers they fund must adhere to.
3. *Researchers* design research projects and work directly with participants and communities during implementation. It is normally they who are responsible for ethics dumping, whether deliberate or inadvertent.
4. Many studies involve human *research participants* who are directly affected by the research. As ethics dumping can also affect animals and the environment, groups working to defend them against unethical treatment could count as advocates – that is, persons who act on behalf of other entities. The same applies to nongovernmental organizations (NGOs) or think-tanks that promote the interests of those who cannot defend themselves against exploitation, or who struggle to do so. Hence, these groups are included in the list of stakeholders.
5. Negative impacts from unethical research conduct can extend beyond research participants and cause harm to *community members*. In genetic research, for instance, research results are likely to be relevant to close family members who were not involved in the study (Gallo et al. 2009). Similarly, a research participant might divulge valuable traditional knowledge held by a community, which cannot be used ethically (or even, sometimes, legally) by the researcher unless s/he has also engaged the wider community (Wynberg et al. 2009).
6. The final group that can count as a major stakeholder in research consists of *research ethics committees*, which review and approve research proposals on behalf of funders or research institutions. This is especially important when tackling ethics dumping, as the role of research ethics committees is to safeguard the rights and welfare of those involved in research (Levine 2004: 2312).

Figure 6.2 graphically depicts the main parties involved in research, and therefore represents the research stakeholders in the fight against ethics dumping.

There were budget-holding representatives from each of these research stakeholder groups in the TRUST project consortium (see Table 6.1). This meant that even before outward engagement to draft the GCC, a lot of information could be generated internally.

As Table 6.1 shows, considerable expertise from different stakeholder perspectives was available internally. In addition, input was sought from external experts, who engaged through four channels, facilitated by six enablers, to participate in the development of a range of project outputs, one of which was the GCC. This approach is detailed in Figure 6.3 (Dammann and Cavallaro 2017).

Fig. 6.2 Research stakeholders

Table 6.1 Main research stakeholders involved as TRUST budget holders

Stakeholder type	TRUST partner
Research policymakers	The United Nations Educational, Scientific and Cultural Organization (UNESCO) promotes policy frameworks for research through drafting internationally focused guidelines on scientific co-operation. The "Institut national de la santé et de la recherche médicale"® (Inserm) holds responsibility for the strategic, scientific and operational coordination of French biomedical research.
Research funders	The European and Developing Countries Clinical Trials Partnership (EDCTP) funds research for the prevention and treatment of poverty-related infectious diseases in sub-Saharan Africa. The Swiss-based Global Values Alliance (GVA) is a foundation that focuses on engagement with pharmaceutical industry partners as its main role in the TRUST project
Researchers	The Centre for Professional Ethics at the University of Central Lancashire is one of the oldest research-only ethics centres in Europe, with specialist expertise in global justice issues. The Bio-Economy Research Chair at the University of Cape Town and her team focus on engagement with communities, indigenous knowledge holders and policymakers to ensure environmentally sustainable poverty reduction. The Law School of the University of the Witwatersrand, Johannesburg, contributed specialist human rights and legal frameworks expertise.
Research participants	Many of the individuals involved in drafting the GCC have previously been research participants. Of particular importance in this process were the indigenous peoples and sex worker representatives who were involved through SASI and PHDA (see "Research communities" below).
Research advocates (support organizations)	The Council on Health Research for Development (COHRED) helps promote the health and development of populations in low- and middle- income countries. Action Contre La Faim (ACF) is recognized as one of the leading organizations in the fight against hunger worldwide. ACF undertakes its own research on highly vulnerable populations.

Table 6.1 (continued)

Stakeholder type	TRUST partner
Research communities	The South African San Institute (SASI) is dedicated to serving the San communities of southern Africa through legal, advocacy, socio-anthropological and related services. Partners for Health and Development in Africa (PHDA) supports female and male sex workers in the low socio-economic strata who reside in the informal settlements of Nairobi.
Research ethics reviewers	The Forum for Ethics Review Committees in India (FERCI) promotes the effective implementation of the ethical review of biomedical research studies in India.

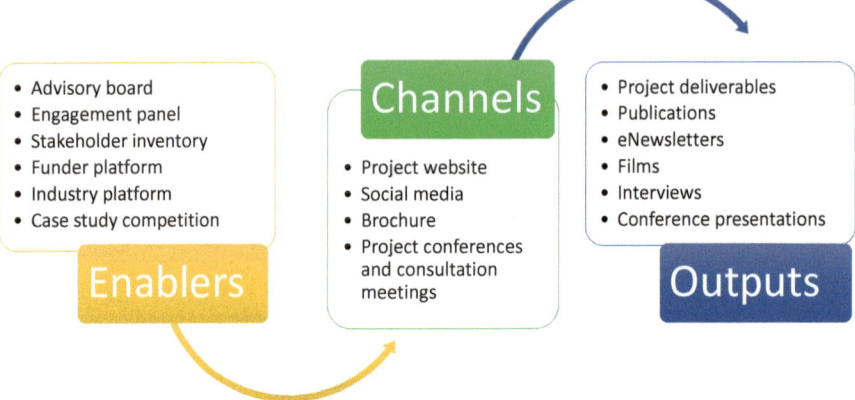

Fig. 6.3 TRUST communication and engagement strategy

External engagement prior to drafting the GCC was extensive. The elements of the communication and engagement strategy that were specifically relevant to the drafting of the GCC are described in more detail below, namely:

- A case study competition
- Project conferences and consultation meetings
- Funder and industry platforms

The Case Study Competition

As the GCC was designed to address the risks of ethics dumping, it was essential to analyse as many actual examples of ethics dumping as possible. To reach out to stakeholders who were not connected with existing networks, a competition to describe cases of ethics dumping was launched in 2016. Applicants from around the world were invited to submit short abstracts of ethics dumping cases, which involved research undertaken in low- and middle- income countries (LMICs) conducted by researchers, sponsors or funders from high-income countries (HICs). Cases could report on events related to any research field. A judging committee selected by TRUST ranked the best ten abstracts, and shortlisted applicants were invited to submit a full case study. Rewards for the authors of the best five cases were €2,000 each and €1,000 each for five runners-up. Following peer review, eight full-length case studies were selected for inclusion in *Ethics Dumping: Case Studies from North-South Research Collaborations* (Schroeder et al. 2018) or as learning materials for the GCC website (http://www.globalcodeofconduct.org). This mechanism expanded the material available for the development of the risk matrix considerably.

Meetings and Platforms: Reaching the Right Delegates

"Around the world, millions of meetings are being held every day – most of them unproductive" (Koshy et al. 2017). In "Not Another Meeting!" Rogelberg et al. (2006) establish that perceived meeting effectiveness has a strong, direct relationship with positive job attitudes and wellbeing at work. In the literature, meeting efficiency is linked to questions such as, "Is a meeting necessary?" (Koshy et al. 2017), or "Is a meeting the most cost-effective way of obtaining an outcome?" (Rogelberg et al. 2006). In addition, advice is given on how to make meetings more efficient, such as, "Prepare the agenda in advance," and "Start with the most strategic items," (Rogelberg et al. 2006).

For the TRUST consortium, the most important question ahead of all major consultation meetings was: "Who are the external delegates?" On the one hand, are they senior and/or from influential institutions? Or, on the other hand, do they have first-hand experience of ethics dumping? In other words, the consortium aimed for senior decision-maker representation as well as vulnerable population representation. To give an example of the former, Table 6.2 shows the funders and companies which were represented at the funder and industry consultation. The consultation with vulnerable populations on engagement with research participants will be described below.

As noted above, millions of meetings are held every day around the world, and many of them affect job satisfaction and wellbeing negatively. How, then, could a meeting hosted on behalf of a three-year research project achieve such impressive representation? There were four reasons for this success:

1. A convincing justification for the meeting secured a Wellcome Trust venue in central London. The Wellcome Trust is the largest private funder of medical research globally (Jack 2012), with a very high standing in research circles.

Table 6.2 Funders and industry members represented at consultation meeting, London 2017

Funders	Industry members
Wellcome Trust	European Federation of Pharmaceutical Industries and
European Commission.	Associations (EFPIA)
Medical Research Council	Sanofi
UKRI	Roche
World Health Organization TDR	Novartis
Calouste Gulbenkian Foundation	GlaxoSmithKline
Global Forum on Bioethics in	Boehringer Ingelheim
Research	

2. Invitations to funders were issued by the European and Developing Countries Clinical Trials Partnership (EDCTP), the high-profile funding institution represented on the TRUST consortium.
3. Invitations to industry were issued by Professor Klaus Leisinger, a member of the TRUST consortium and former personal adviser on corporate responsibility to UN Secretary-Generals Kofi Annan and Ban Ki-moon.
4. Most important to the success, however, were the prior activities of the funder and industry platforms.

TRUST's Funder Platform was established by the EDCTP in 2016 via the following steps:

- Development of an inventory of major funders of basic and operational/implementation research around the world.
- Obtaining the names of those whose expertise included research ethics in LMICs, ethics committees in LMICs, adherence to ethical standards, ethics codes and best practice.
- Introduction of the project and its aims (e.g. the GCC) in personal, tailored communications to the experts identified.
- Personal invitations to the funder consultation meeting in London.
- Personal, tailored follow-up communication after the meeting.

Policymakers at the highest UN level have selected the pharmaceutical sector for special responsibilities towards LMICs, and therefore this sector was chosen for engagement activities with industry in TRUST. Both the Millennium Development Goals[4] and the Sustainable Development Goals[5] call on the pharmaceutical industry for special assistance to LMICs. In addition, the sector suffers from considerable mistrust among the general population with regard to international collaborative research and the potential exploitation of LMICs (Kessel 2014).

TRUST's Industry Platform was established in 2016 by Professor Leisinger in the same manner as the Funder Platform, with the following addition:

- Webinars and personal meetings were organized through EFPIA, at which Professor Leisinger explained the ambitions of the TRUST project and the need for pharmaceutical companies to contribute.

Through this extensive preparation, it was possible to introduce draft ideas for the GCC at a high-profile and well-attended consultation meeting in London in June 2017, and to engage delegates sufficiently to secure further input, nine months later, on the first semipublic draft of the GCC.

Figure 6.4 gives an overview of the various project conferences and consultation meetings at which information was sought to counter ethics dumping.[6]

---

[4] Goal 8, Target 4: "in cooperation with pharmaceutical companies, provide access to affordable essential drugs in developing countries" (United Nations ndb).

[5] Goal 3, Target 3.8: "Achieve universal ... access to safe, effective, quality and affordable essential medicines and vaccines for all" (United Nations nda).

[6] The meetings had a range of other purposes, many of which are not relevant here and are therefore not included in this summary.

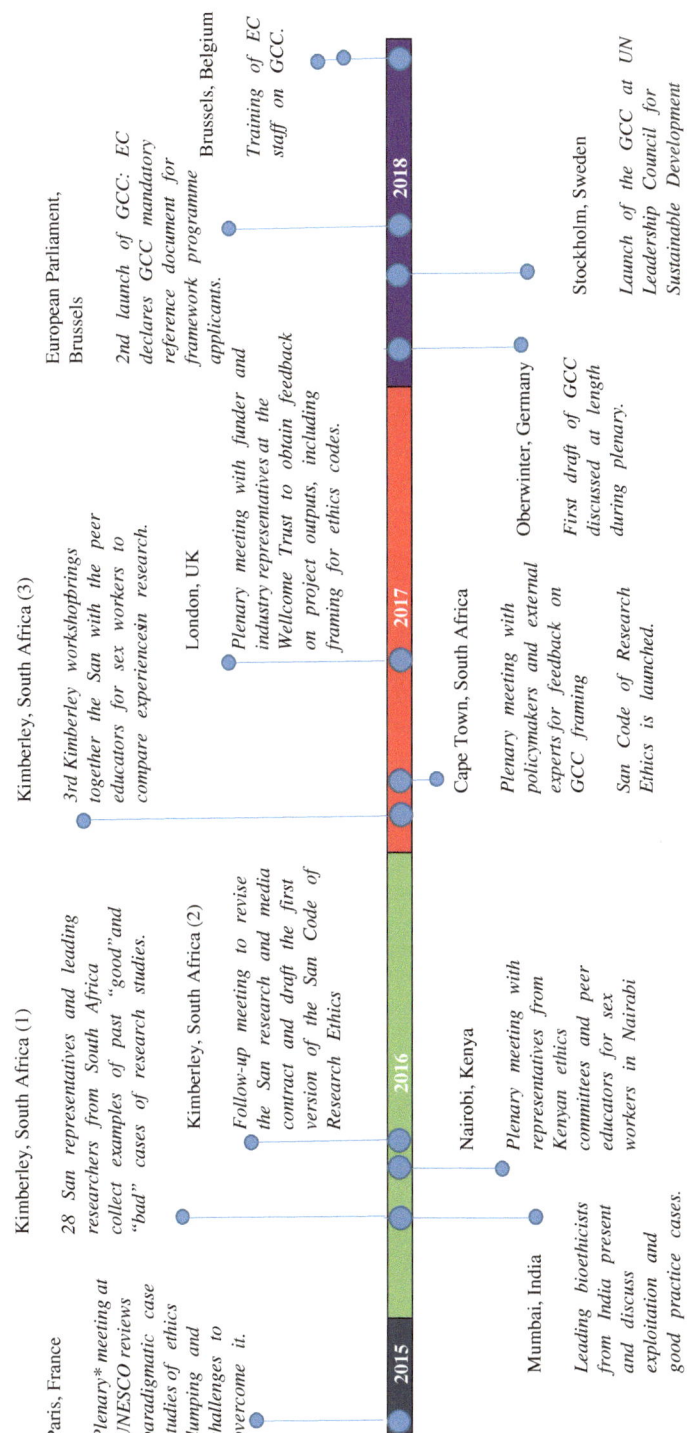

* The term "plenary" refers to all meetings that the entire TRUST consortium attended.

Fig. 6.4 Consultations for and launch of the GCC

Returning to the six research stakeholder groups identified earlier, the next section details how each group was reached through project conferences and consultation meetings.

External Engagement with Research Policymakers

National research foundations, research councils and government ministries guide the strategic direction of research. Representatives from all of these groups attended the TRUST plenary in Cape Town in 2017, in particular senior representatives from the following national bodies:

- The South African Department of Science and Technology
- The South African Department of Environmental Affairs
- The South African National Research Foundation
- The Zimbabwean Agricultural Research Council

To give an example of input, articles 1[7] and 4[8] of the GCC are directly linked to input from research policymakers. Dr Isaiah Mharapara from the Zimbabwean Agricultural Research Council argued that agricultural research in Africa had largely been based on foreign principles, meaning that the continent's own crops, fruits, insects, fish and animals had been ignored. Through the historical introduction of Western agricultural systems and cash crops such as tobacco, as well as genetically engineered crops, Africa had failed to develop agricultural solutions adapted to local conditions. According to Dr Mharapara, a lack of financial resources meant that African nations had been, and still were, vulnerable to exploitation by foreign researchers. This had resulted in damage to ecological systems, the loss of soils, fertility, biodiversity and natural resilience, and the erosion of indigenous knowledge. He advocated inclusive, consultative, robust and agreed processes to establish equitable research partnerships (Van Niekerk et al. 2017).

External Engagement with Research Funders

Estimates for research and development expenditure in the European Union in 2016 indicate that 56.6% of all such expenditure comes from the business sector, 30.9% from the government sector and the remainder mostly from charitable foundations (Eurostat 2018). TRUST's main consultation workshop for research funders was held in London in 2017 and involved all three sectors: public funders, private

---

[7] Local relevance of research … should be determined in collaboration with local partners.

[8] Local researchers should be included, wherever possible, throughout the research process, including in study design, study implementation, data ownership, intellectual property and authorship of publications.

Table 6.3 Good practice input from funders and industry with GCC output

Good practice	Relevant GCC article
Ensuring double ethics review	Article 10: Local ethics review should be sought wherever possible.
Community engagement	Article 2: Local communities and research participants should be included throughout the research process.
Clear roles and responsibilities	Article 20: A clear understanding should be reached among collaborators with regard to their roles, responsibilities and conduct throughout the research cycle.

funders and charitable funders of research (see Table 6.2). The three main good practice elements[9] raised by funders and industry to stop ethics dumping are listed in Table 6.3, with their corresponding GCC articles (Singh and Makanga 2017).

As already indicated, engagement with research funders was not restricted to one meeting, but took place over approximately two years via the funder and industry platforms described above. Additionally, the first draft of the GCC was distributed to all members of the platforms nine months after the workshop. Both groups provided further comments on the draft.

External Engagement with Researchers

The consortium that drafted the GCC represented a wide range of academic disciplines, namely ethics, medicine, economics, bioethics, law, social psychology, sociology, psychology, gender studies, chemistry, social sciences, psychiatry, biology, zoology, veterinary medicine, political science and management. The multidisciplinary nature of the consortium's expertise enabled broad engagement with the wider academic community. For example, academic presentations that included the GCC were delivered in Belgium, China, Congo, Cyprus, Germany, India, Kenya, Latvia, the Netherlands, the Philippines, Portugal, Rwanda, South Africa, Sweden, Taiwan, Uganda, Vatican City, the UK and the USA (Dammann and Schroeder 2018).

The feedback from researchers was essentially threefold. First, researchers were interested in the potential "grey areas" of ethics dumping. A question in this context, asked on many occasions by different audiences, was: "If a particular research study has no real local relevance to LMICs, and the research money spent by well-intentioned researchers from HICs (who genuinely believe that they are improving the world) is in fact being wasted, does that count as ethics dumping?" A case in

---

[9] A fourth good practice element that was emphasized at the funder and industry workshop was the provision of post-trial access to successfully marketed drugs. This requirement was not included in the GCC for two main reasons. First, the GCC was designed to be applicable to all disciplines, and hence articles with limited applicability were avoided. Second, post-trial access is clearly indicated in existing guidelines, in particular in the Declaration of Helsinki (WMA 2013). A good practice example of post-trial access was included in a TRUST special symposium on industry obligations towards LMICs (see Kelman et al. 2019).

point has been described by Van Niekerk and Wynberg (2018). Northern researchers were working on a genetically modified banana with a purportedly enhanced vitamin content, the ultimate aim being to alleviate nutritional deficiencies in Uganda. However, it turned out that existing local varieties provided a higher vitamin content than the envisaged GM variety. The question of whether that should count as ethics dumping goes back to a long-standing dilemma for ethics committees: Is bad science bad ethics? (Levine 2004) Wasteful research cannot be put into the same category as the wilful exploitation of lower regulatory standards to exploit research populations in LMICs for individual gain. But at the same time, with such a pressing need for innovative solutions to LMIC problems, the violation of article 1 of the GCC (local relevance of research) and the *avoidable* waste of limited funding resources must count as unethical.

A second recurring response to the GCC from researchers has been that "everyone loved our values".[10] Audiences in HICs – England, for instance – even asked whether they could use the four values in national research in their own countries. Hence, rather than seeing the values as solely applicable when there are vast power differentials between researchers and research participants (as between HICs and LMICs), they were keen to use them in *any* research.

A third recurring issue for researchers has been the following: "We appreciate that the code is short and accessible, but wouldn't a longer, more detailed code give more support to early career researchers?" The TRUST consortium agreed upon a concise code because it is vital that the demands of substance for each article be clear and straightforward, while the process demands remain flexible. Let us take article 1 as an example:

The *substance element* of article 1 is: "Local relevance of research is essential". Further information would not be helpful to early career (or any other) researchers.

The *process element* of article 1 is: "[Local relevance] should be determined in collaboration with local partners." This could only be set out in more detail if there were a single process that would fit every situation – and that is not the case. What an equitable process for determining research goals should look like in an international collaborative research project is one of the things that need to be agreed on within the process of that project. Hence, prescriptive details would have been counterproductive to the very spirit of the article.

Instead of attempting to formulate a range of possibilities to fill the process elements with substance, we opted to provide educational material to support the GCC online,[11] because any process requirements are best agreed between the relevant partners rather than imposed prescriptively by code drafters. Hence, our educational materials future-proof the GCC, as they can be updated in real time for use by early career (or any other) researchers, and, unlike the GCC itself, they are not mandatory.

---

[10] Personal communication from Dr Vasantha Muthuswamy, a TRUST project team member, after a GCC presentation in Taiwan.

[11] http://www.globalcodeofconduct.org/

Engagement with Research Participants and Research Communities

The inclusion of the perspectives of research participants and research communities who are vulnerable to exploitation, and therefore to ethics dumping, was essential to our bottom-up approach. It is also the ethical approach, as stipulated in article 2 of the GCC:

> Local communities and research participants should be included throughout the research process, wherever possible, from planning through to post-study feedback and evaluation, to ensure that their perspectives are fairly represented. This approach represents Good Participatory Practice.

Two NGO partners in the TRUST project were tasked specifically with ensuring that the voices of vulnerable populations were heard and acted upon. First, the South African San Institute (SASI) made the inclusion of indigenous peoples from South Africa possible. While San leaders and representatives were involved in all the work of the TRUST project, including the drafting of the GCC, the full impact of their contribution is best understood through the account in Chapter 7 of this book of the development of the San Code of Research Ethics. Second, Partners for Health and Development in Africa (PHDA) made the inclusion of sex workers from the Majengo area of Nairobi possible. At this point we will focus on their involvement in order to illustrate the bottom-up approach of the GCC drafting process.

PHDA is a nonprofit organization that undertakes work in the fields of health and development in Kenya. Its mission is to increase access to health for disadvantaged communities in Africa by strengthening health systems, research, programme development and partnerships. PHDA's programmes are implemented by a collaborative group of scientists and public health professionals from the University of Manitoba (Canada), the University of Nairobi and the government of Kenya. Its work focuses mainly on HIV prevention, treatment and care, research, capacity-building and training.

The Sex Workers Outreach Programme (SWOP) is a PHDA initiative that undertakes active community engagement and provides clinical and preventative services to 33,000 sex workers residing in Nairobi. These sex workers would otherwise find access to medical services in public health facilities extremely limited due to stigma and discrimination. Those enrolled in the sex workers cohort for HIV prevention services are free to volunteer for available research studies after providing informed consent. Most studies are on the epidemiology of sexually transmitted diseases, and on host genetic factors that influence infectivity and disease progression.

Given that sex work is illegal in Kenya, we cannot assign input to specific, named individuals here. Suffice to say that the personal contributions of courageous and admirable sex workers, both female and male, provided the TRUST team not only with practical advice that took shape in specific articles of the GCC, but also with inspiration. Table 6.4 presents two examples of issues raised by the Nairobi sex workers (Chatfield et al. 2016a) that were implemented in the GCC.

Table 6.4 Input from sex workers and GCC connection

Issues raised by sex workers	Relevant GCC Article
"We need feedback to the community from the research in simple and non-scientific language. Some results have been shared with us in the past, but I did not know what they meant. Do not give us results in scientific language. It puts us at risk if we do not understand the results. … Come back with the results and tell us how we can make our lives better."	Article 3: Feedback about the findings of the research must be given to local communities and research participants. It should be provided in a way that is meaningful, appropriate and readily comprehended.
"We know that the samples that are collected from us are sometimes sent to other countries. What happens to them? In my culture – if my blood is taken, it must come back to me and I bury it. … [L]ocal and cultural values should be taken into account."	Article 8: Potential cultural sensitivities should be explored in advance of research with local communities, research participants and local researchers to avoid violating customary practices. … If researchers from high-income settings cannot agree on a way of undertaking the research that is acceptable to local stakeholders, it should not take place.

The main message that the TRUST team has been promoting since the meetings with the Nairobi sex workers has been: "Let representatives of vulnerable populations speak for themselves" (Schroeder and Tavlaki 2018). As a result, a former sex worker from Nairobi brought the demands of her community to the European Parliament to great acclaim (TRUST 2018).

Advocate Voices for Animals

A senior veterinarian, Professor David Morton, was involved throughout the TRUST project as an adviser. At a plenary meeting in Cape Town, he described how animals had no voice and therefore no choice about involvement in research. He asked: "Who consents on behalf of animals?"

There are currently no globally agreed ethical standards for research involving animal experimentation, and regulation varies from country to country. In the EU, animal experiments are governed by Directive 2010/63/EU, known as the Animal Experiments Directive, which stipulates measures that must be taken to replace, reduce and refine (the "Three Rs"[12]) the use of animals in scientific research. Among other requirements, it lays down minimum standards for housing and care and regu-

---

[12] The "Three Rs" are the underpinning requirements of most policies and regulations in animal research:

  → Replacement: Methods that avoid or replace the use of animals.
  → Reduction: Methods that minimize the number of animals used per experiment.
  → Refinement: Methods that minimize suffering and improve welfare.

lates the use of animals through systematic project evaluation that requires the assessment of pain, suffering, distress and lasting harm caused to the animals.

Some researchers knowingly exploit variations in standards and opt to conduct animal studies in LMICs because it is cheaper and/or because regulation is less strict than in HICs (Morton and Chatfield 2018). For example, researchers might conduct experiments on non-human primates in an LMIC setting that would be illegal in their HIC home country.

For research collaborations between groups in different countries, partners may find that they are confronted with different ethical standards for animal experimentation. In such cases ethical standards should comply with the highest ethical standards rather than be adjusted to the lowest common denominator. Hence the GCC states that standards for animal research in international collaborative research must comply with those that are more demanding and protective of animal welfare (article 17).

External Engagement with Research Ethics Committees

The main engagement meetings with research ethics reviewers and chairs of research ethics committees took place in India (2016) and Kenya (2017).

The Forum for Ethics Review Committees in India is led by Dr Vasantha Muthuswamy, who was responsible for issuing the Indian Council of Medical Research's *Ethical Guidelines for Biomedical Research on Human Subjects* in 2000 and the revised version, *Ethical Guidelines for Biomedical Research on Human Participants*, in 2006, and also contributed to the most recently launched version in 2017. At her invitation, 30 leading bioethicists from India came together with guests from Europe in a two-day workshop in Mumbai in 2016. This workshop, which was attended by many senior research ethics committee chairs and members, was an important fact-finding mission in the early stages of the project. Cases of exploitation were collated and good practice in research involving LMICs discussed (Chatfield et al. 2016b).

Further in-depth consultation with ethics committee chairs formed part of a plenary workshop in Nairobi in 2017. TRUST received valuable input from three esteemed ethicists: Professor Elizabeth Bukusi (Deputy Director Research and Development, Kenya Medical Research Institute), Professor Anastasia Guantai (Kenyatta National Hospital) and Professor Kirana Bhatt (Chair of the National Bioethics Committee, University of Nairobi), who shared their respective experiences, concerns and insights. Table 6.5 summarizes some of the issues raised (Chatfield et al. 2016b) and their relationship to the final GCC.

Table 6.5 Input from research ethics committee chairs and GCC output

Input	Relevant GCC article
Intellectual property rights are often held only in the North.	Article 4: Local researchers should be included, wherever possible, throughout the research process, including in study design, study implementation, data ownership, intellectual property and authorship of publications.
Attempts at gaining ethics approvals can be extremely late.	Article 11: Researchers from high-income settings should show respect to host country research ethics committees.
LMIC partners' tasks are restricted to obtaining data.	Article 4 (see above) Article 20: A clear understanding should be reached among collaborators with regard to their roles, responsibilities and conduct throughout the research cycle.
Why does biological material need to be shipped abroad?	Article 5: Access by researchers to any … human biological materials … should be subject to the free and prior informed consent of the owners or custodians. Formal agreements should govern the transfer of any material or knowledge to researchers, on terms that are co-developed with resource custodians or knowledge holders.

Analysis of Existing Guidelines

We have summarized above the extensive consultation activities of the TRUST project prior to the actual drafting of the GCC. Aside from these valuable contributions, it was also vital that the GCC should not set out to "reinvent the wheel". Given the vast number of existing guidelines, and the significant expertise that went into drafting them, it was important for us to link the GCC to those existing guidelines so as to produce something that did not replicate earlier work, but rather complemented it.

Research ethics committees have been in operation since the 1960s (Levine 2004). The earliest codes of research ethics are even older (Levine 2004). Yet the ethics dumping cases identified as part of the fact-finding mission for the GCC occurred mostly in the 2010s (Schroeder et al. 2018), more than 50 years after the first codes and committees became operational. There are many reasons why ethics dumping in research persists, one of which is that the constraints on research ethics committees in LMICs make them vulnerable to exploitation, often across the North–South global divide. Some such constraints are summarized in Table 6.6 (modified from Nyika et al. 2009).

A new code of ethics cannot, by itself, resolve these issues, for instance the lack of resources to fund effective ethics review. However, it can tailor requirements to LMIC needs. To do so most effectively, the TRUST team decided that it would *not* base the drafting process on existing ethics guidelines, because, as is also noted in Chapter 4, the history of research ethics review is heavily built on the United States experience and context and focused on medical research (see also Levine 2004). To avoid any potential bias, the following approach was agreed upon. First, the exercise of identifying ethics dumping risks was to be carried out without reference to existing ethics guidelines. Second, only once the risks had been identified through fact-

Table 6.6 Constraints on African research ethics committees

Insufficiency of resources	No or poor support by the hosting institution
Lack of or insufficient expertise on ethical review	Not completely independent
Pressure from researchers	Pressure from sponsors
Lack of active or consistent participation of members	Unequal treatment of applicants in review
Lack of recognition of the importance of REC functions	

finding missions and consultations would existing ethics codes be analysed for their relevance to mitigating the identified risks.

After the risk research was concluded, a meticulous analysis of all identified risks for ethics dumping (chapter 5) was undertaken. All 88 risks were mapped onto existing ethics codes. For instance, Risk 8 was identified as "Poor representation of Southern (host) partners on research teams, e.g. responsible for menial tasks only or not acknowledged or represented appropriately in publications" (Singh and Schroeder 2017). The mapping exercise discovered that the best match to mitigate this risk was the Montreal Statement on Research Integrity in Cross-Boundary Research Collaborations (Singh and Schroeder 2017), which notes:

> Collaborating partners should come to an agreement, at the outset and later as needed, on how publication and other dissemination decisions will be made and on standards for authorship and acknowledgement of joint research products. The contributions of all partners, especially junior partners, should receive full and appropriate recognition. (Montreal Statement 2013)

Through this exercise, it was possible to arrive at guidance to counter ethics dumping, while simultaneously identifying which risks, if any, were not covered by *any* notable guidance. For instance, the lack of guidance on risk management approaches to biosafety and biosecurity was discovered in this way (Singh and Schroeder 2017), leading to article 18[13] of the GCC.

Once the fact-finding, consultations and analysis had been done, the drafting process began.

Drafting Process

The drafting committee of the GCC consisted of four people, following Michael Davis's (2007) advice:

> Keep the drafting committee small. Preparing a *first* draft of a code is not an activity made lighter "by many hands". It is more like the soup that "too many cooks" spoil."

---

[13] In situations where environmental protection and biorisk-related regulations are inadequate or non-existent in the local setting compared with the country of origin of the researcher, research should always be undertaken in line with the higher standards of environmental protection.

Table 6.7 GCC drafting committee

Drafter	Region of origin	Focus	Background	Roles
Schroeder	North	All engagement and fact-finding	Philosophy, politics, economics	Full first draft
Chennells	South	Vulnerable populations	Law	Drafting articles to protect vulnerable populations
Chatfield	North	Risks	Social science, philosophy	Redrafting to ensure all risks were covered
Singh	South	Existing guidelines	Public health	Redrafting with a focus on existing guidelines

Table 6.7 shows the configuration of the four-person drafting committee. The emphasis was on 50% North and 50% South membership, taking into account the expertise needed and relevant background.

The four-values approach, featuring fairness, respect, care and honesty, had been adopted by the consortium at an earlier stage (chapter 3). Based on these values and the inputs into the GCC from consultations and fact-finding activities over two years (see Fig. 6.1), the lead author, Professor Doris Schroeder, drafted the first version, which contained 20 articles, three fewer than the final version.

For instance, article 15 on the risks of stigmatization, incrimination, discrimination and indeterminate personal risk was added during the peer review process by Professor Morton, the veterinary expert in the consortium.

Professor Schroeder's first draft was refined considerably by the social science and risk expert on the drafting committee, Dr Kate Chatfield. Dr Roger Chennells, the expert on involving vulnerable populations in research, who also provided a legal perspective, drafted his own articles on the prevention of ethics dumping. Many of these addressed the same issues identified by the lead author, but now illuminated by a legal reading. For example, article 20 on the clear understanding of roles and responsibilities was an important addition. Finally, the expert on existing guidelines, Dr Michelle Singh, checked the draft code for oversights relevant to countering ethics dumping and also, for example, added article 7 on the importance of compensating local support systems.

The first full draft agreed by the four-person committee then went through a rigorous internal peer review process in the consortium, including detailed discussions at a plenary meeting in Germany in February 2018. Each draft article was analysed in depth. Changes at this stage included:

- A different order to demonstrate importance through emphasis: for example, the assertion that the local relevance of research is essential became article 1.
- Different wording: for example, the word "gatekeeper' in article 7 was replaced with "local coordinator", and "community approval" in article 9 was changed to "community assent".
- Regrouping to connect with a different value: for example, local ethics review was moved from the "fairness" section to "respect".

- Combining articles: for example, showing respect to local ethics committees and producing documentation in line with local requirements were combined in the final article 11.

Some issues could not be solved at the meeting and individual experts were tasked with presenting solutions. For instance, Professor Rachel Wynberg from the University of Cape Town, a world-leading expert on benefit sharing, was asked to ensure that the benefit-sharing articles of the GCC (articles 5 and 6) were in line with the most recent legal instruments and phrased as accessibly as possible.

Article 14 was debated with particular gusto. Part of the group wanted to prevent *any* research that was prohibited in HICs from taking place in LMICs, whereas others wanted to leave room for justifiable exceptions. The final version of the article was agreed through consensus-building suggestions from Professors Carel IJsselmuiden and Klaus Leisinger, which enabled permissible exceptions with a "comply or explain" proviso.

When an internal consensus was reached on the entire text, bearing in mind that "internal" involved 13 partner organizations from around the world, the agreed draft was released to the previously engaged external stakeholders, in line with further advice from Michael Davis (2007):

> Make the procedure as open as possible once there is a first draft. The openness … protects the drafting committee not only from the eccentricities of those outside the committee but from the tendency of drafting committees to forget practical constraints.

All articles were accordingly submitted to a much broader peer review by those who had previously taken part in consultations, and their comments were obtained. Many of these comments were acted upon, including:

- An industry representative's suggestion to remove the following closing sentence from the draft article on community assent: "Developing personal, long-standing relationships with local communities produces the bedrock of respect." This was on the basis that the statement did not apply to *all* international collaborative research, and also that no other GCC article had such a commentary.
- The request by several external stakeholders from various backgrounds that the qualification "wherever possible" be added to articles 2, 4 and 10, in order to be more realistic.
- The proposal that "vehicle drivers" be removed from the examples of local research support systems given in article 7.

After this extensive peer review, three further activities were undertaken. First, the final draft was reworked by a professional editor (Paul Wise in South Africa) to achieve the clearest, most precise and most accessible language. Second, the code was professionally designed for publication (CD Marketing Ltd, UK). Third, funding was obtained (from the University of Central Lancashire, UK, and the EDCTP) to translate the code into Russian, French, Spanish, German, Portuguese, Mandarin, Japanese, and Hindi.[14] Arabic and isiXhosa translations are in progress as this book goes to press.

---

[14] The translations can be found at http://www.globalcodeofconduct.org/open-to-the-world/

Early Adopters and Conclusion

After an intensive and scrupulous development process, the GCC was launched at two high-profile events. First, at a meeting of the UN Leadership Council for Sustainable Development in Stockholm, Sweden, in May 2018, and then at the European Parliament in June 2018. The GCC had been under examination by the ethics and integrity sector and the legal services of the EC since March 2018. This allowed Dr Wolfgang Burtscher, Deputy Director-General for Research and Innovation of the EC, to announce the big news at the European Parliament event: the GCC would be a mandatory reference document for the framework programmes that fund European research. What this means was expressed succinctly in a *Nature* article.

> Ron Iphofen, an adviser on research ethics to the European Commission, believes the code will have a profound impact on how funding proposals to the EU are designed and reviewed. "I could envisage reviewers now looking suspiciously at any application for funds that entailed research by wealthy nations on the less wealthy that did not mention the code," he says. (Nordling 2018)

Two months later, in August 2018, the EDCTP announced that henceforth its applicants would be required to comply with the GCC. In April 2019, the Senate of the University of Cape Town (UCT) adopted the GCC as the first university globally to ensure that UCT researchers maintain the highest ethical standards.

The groundwork for developing the GCC included a broad collation of ethics dumping case studies, as well as good practice examples from international collaborative research, and extensive consultation with representatives from a range of stakeholder groups: research policymakers, research funders including private industry, researchers, research communities, research ethics committees and, most importantly, vulnerable research participants and those who support them. Building on 88 generic risks identified in the fact-finding and consultation phases of the TRUST project, and taking existing guidelines into account, a code was built which will provide guidance across all research disciplines. It focuses on research collaborations between LMIC and HIC partners, which often involve considerable imbalances of power, resources and knowledge. The GCC is presented in 23 clear, short statements in order to achieve the highest possible cross-cultural accessibility for researchers, funders and vulnerable populations alike.

Those who apply the GCC are demonstrating that they oppose double standards in research and support long-term equitable research relationships between partners from LMIC and HIC settings, based upon the values of fairness, respect, care and honesty.

References

Bassler A, Brasier K, Fogel N, Taverno R (2008) Developing effective citizen engagement: a how-to guide for community leaders. Center for Rural Pennsylvania, Harrisburg PA. http://www.rural.palegislature.us/effective_citizen_engagement.pdf

Burtscher W (2018) TRUST Global Code of Conduct to be a reference document applied by all research projects applying for H2020 funding. TRUST eNewsletter Issue 5. http://www.global-codeofconduct.org/wp-content/uploads/2018/12/TRUSTNewsletter_2018_Issue5.pdf

Chatfield K, Schroeder D, Kimani J (2016a) Vulnerable populations in North-South collaborative research: Nairobi plenary 2016. A report for TRUST. http://trust-project.eu/wp-content/uploads/2016/11/Meeting-Report-TRUST-Nairobi-Final.pdf

Chatfield K, Schroeder D, Muthuswamy V (2016b) Mumbai case studies meeting. A report for TRUST. http://trust-project.eu/wp-content/uploads/2016/06/Mumbai-Case-Studies-Workshop.pdf

Cook WK (2008) Integrating research and action: a systematic review of community-based participatory research to address health disparities in environmental and occupational health in the United States. Journal of Epidemiology and Community Health 62(8):668–676. https://doi.org/10.1136/jech.2007.067645

Dammann J, Cavallaro F (2017) First engagement report. A report for TRUST. http://trust-project.eu/wp-content/uploads/2016/03/TRUST-1st-Engagement-Report_Final.pdf

Dammann J, Schroeder D (2018) Second engagement report. A report for TRUST. http://trust-project.eu/wp-content/uploads/2018/12/TRUST-2nd-Engagement-Report-Final.pdf

Davis M (2007) Eighteen rules for writing a code of professional ethics. Science and Engineering Ethics 13(2):171–189.

Dunn A (2011) Community engagement: under the microscope. Wellcome Trust, London. https://wellcome.ac.uk/sites/default/files/wtvm054326_0.pdf

European Commission (2013) Declarations of the Commission (framework programme) 2013/C 373/02. Official Journal of the European Union 20 December. http://ec.europa.eu/research/participants/data/ref/h2020/legal_basis/fp/h2020-eu-decl-fp_en.pdf

Eurostat (2018) Intramural R&D expenditure (GERD) by source of funds. https://ec.europa.eu/eurostat/web/products-datasets/product?code=rd_e_gerdfund

Gallo AM., Angst DB, Knafl KA (2009) Disclosure of genetic information within families. The American Journal of Nursing 109(4):65–69. http://www.ncbi.nlm.nih.gov/pubmed/19325321

Hebert JR, Brandt HM, Armstead CA, Adams SA, Steck SE (2009) Interdisciplinary, translational, and community-based participatory research: finding a common language to improve cancer research. Cancer Epidemiology, Biomarkers & Prevention 18(4):1213–1217. https://doi.org/10.1158/1055-9965.EPI-08-1166

Horizon 2020 (nd) Evaluation of proposals. Research and Innovation, European Commission. http://ec.europa.eu/research/participants/docs/h2020-funding-guide/grants/from-evaluation-to-grant-signature/evaluation-of-proposals_en.htm

Jack A (2012) Wellcome challenges science journals. Financial Times, 10 April. https://www.ft.com/content/81529c58-8330-11e1-ab78-00144feab49a

Kelman A, Kang A, Crawford B (2019) Continued access to investigational medicinal products for clinical trial participants: an industry approach. Cambridge Quarterly of Healthcare Ethics 28(1):124–133. https://www.cambridge.org/core/product/identifier/S0963180118000464/type/journal_article

Kessel M (2014) Restoring the pharmaceutical industry's reputation. Nature Biotechnology 32:983–990. https://doi.org/10.1038/nbt.3036

Koshy K, Liu A, Whitehurst K, Gundogan B, Al Omran Y (2017) How to hold an effective meeting. International Journal of Surgery: Oncology 2(5):e22. Available at: http://www.ncbi.nlm.nih.gov/pubmed/29732455

Levine R (2004) Research ethics committees. In: Post S (ed) Encyclopedia of bioethics, 3rd edn. Thomson & Gale, New York, p 2311–2316

Montreal Statement (2013) Montreal statement on research integrity in cross-boundary research collaborations. http://ethics.iit.edu/codes/WCRI%202013.pdf

Morton D, Chatfield K (2018) The use of non-human primates in research. In: Schroeder D, Cook J, Hirsch F, Fenet S, Muthuswamy V (eds) Ethics dumping: case studies from North-South research collaborations. Springer Briefs in Research and Innovation Governance, Berlin, p 81–89

Nordling L (2018) Europe's biggest research fund cracks down on "ethics dumping". Nature 559:17–18. https://www.nature.com/articles/d41586-018-05616-w

Nyika A, Kilama W, Chilengi R, Tangwa G, Tindana P, Ndebele P, Ikingura J (2009) Composition, training needs and independence of ethics review committees across Africa: are the gate-keepers rising to the emerging challenge? Journal of Medical Ethics 35(3):189–193

Rogelberg SG, Leach DJ, Warr PB, Burnfield J L (2006) "Not another meeting!" Are meeting time demands related to employee well-being? Journal of Applied Psychology 91(1):83–96

Schroeder D, Cook J, Hirsch F, Fenet S, Muthuswamy, V (eds) (2018) Ethics dumping: case studies from North-South research collaborations, Springer Briefs in Research and Innovation Governance, Berlin

Schroeder D, Tavlaki E (2018) 2nd pictorial report. A report for TRUST. http://trust-project.eu/wp-content/uploads/2018/12/TRUST-2nd-Pictorial-Report.pdf

Singh M, Makanga M (2017) Funder platform. A report for TRUST. http://trust-project.eu/wp-content/uploads/2017/09/Funder-Platform-Report_20-Sep-2017_funder-approved.pdf

Singh M, Schroeder D (2017) Exploitation risks and research ethics guidelines. A report for TRUST. http://trust-project.eu/wp-content/uploads/2016/03/TRUST-Deliverable-Risks-Principles-Final-for-submission.pdf

TRUST (2018) Strong speech by Nairobi activist in European Parliament. http://trust-project.eu/strong-speech-by-nairobi-activist-in-european-parliament/

United Nations (nda) Goal 3. Sustainable Development Goals. https://www.un.org/sustainabledevelopment/health/

United Nations (ndb) Goal 8. Millennium Development Goals. http://www.un.org/millennium-goals/global.shtml

Van Niekerk J, Wynberg R (2018) Human food trial of a transgenic fruit. In: Schroeder D, Cook J, Hirsch F, Fenet S, Muthuswamy V (eds) Ethics dumping: case studies from North-South research collaborations. Springer Briefs in Research and Innovation Governance, Berlin, p 91–98

Van Niekerk J, Wynberg R, Chatfield K (2017). Cape Town plenary meeting report. TRUST Project. http://trust-project.eu/wp-content/uploads/2017/08/TRUST-Kalk-Bay-2017-Report-Final.pdf

WMA (2013) Declaration of Helsinki. World Medical Association. https://www.wma.net/policies-post/wma-declaration-of-helsinki-ethical-principles-for-medical-research-involving-human-subjects/

Wynberg R, Schroeder D, Chennells R (2009) Indigenous peoples, consent and benefit sharing: lessons from the San-Hoodia case. Springer, Berlin

Youdelis M (2016) "They could take you out for coffee and call it consultation!" The colonial antipolitics of indigenous consultation in Jasper National Park. Environment and Planning A: Economy and Space 48(7):1374–1392

Chapter 7
The San Code of Research Ethics

Abstract The San peoples of southern Africa have been the object of much academic research over centuries. In recent years, San leaders have become increasingly convinced that most academic research on their communities has been neither requested, nor useful, nor protected in any meaningful way. In many cases dissatisfaction, if not actual harm, has been the result. In 2017, the South African San finally published the San Code of Research Ethics, which requires all researchers intending to engage with San communities to commit to four central values, namely fairness, respect, care and honesty, as well as to comply with a simple process of community approval. The code is the first ethics code developed and launched by an indigenous population in Africa. Key to this achievement were: dedicated San leaders of integrity, supportive NGOs, legal assistance and long-term research collaborations with key individuals who undertook fund-raising and provided strategic support.

Keywords San Peoples · Global ethics · Research ethics · Indigenous peoples · Low-and middle-income countries · Ethics dumping

Introduction[1]

The San peoples, widely known as "first" or "indigenous" peoples of southern Africa, have been the object of much academic research over centuries.

In recent years San leaders have become increasingly convinced that most academic research on their communities has been neither requested, nor useful, nor protected in any meaningful way. In many cases dissatisfaction, if not actual harm, has been the result. For instance, a genomics study published in 2010, based on the DNA of four San individuals, included conclusions which San community leaders found "private, pejorative, discriminatory and inappropriate" (Chennells and

---

[1] This chapter is based on a longer, illustrated report (Chennells and Schroeder 2019).

© The Author(s), under exclusive license to Springer Nature Switzerland AG 2019
D. Schroeder et al., *Equitable Research Partnerships*, SpringerBriefs in Research and Innovation Governance, https://doi.org/10.1007/978-3-030-15745-6_7

Steenkamp 2018). Authors of the paper "refused to provide details about the informed consent process .... [and] defended their denial of the right of the San leadership to further information on the grounds that the research project had been fully approved by ethics committees/institutional review boards" (Chennells and Steenkamp 2018).

In March 2017, the South African San launched the San Code of Research Ethics, the first ethics code developed and published by an indigenous community in Africa (Callaway 2017). The code requires all researchers intending to engage with San communities to commit to four central values, namely fairness, respect, care and honesty, as well as to comply with a simple process of community approval.

This chapter introduces the San of southern Africa and the main San support institutions involved in producing the San Code of Research Ethics. It goes on to describe key elements in the development and the launch of the code, namely leaders of integrity, legal support, supportive research collaborations and the process of drafting. Finally, the code is reproduced in full.

The San of Southern Africa

The San peoples of Africa are iconic, widely known as the quintessential hunter-gatherers of Africa and said to be the oldest genetic ancestors of modern humans (Knight et al. 2003). Once ranging over the whole of southern Africa, their numbers have now dwindled to approximately 100,000 San living primarily in Botswana, Namibia and South Africa, with small remnant populations in Angola, Zimbabwe and Zambia (Hitchcock et al. 2006). Although they speak at least seven distinct languages[2] with numerous subdialects, they nevertheless recognize a common cultural identity which is readily identified as a hunter-gatherer heritage, with a shared ancestry also confirmed by genetic research (Soodyall 2006).

Prior to 1990, the San peoples lived typically in extended families and small clans in the remote reaches of South Africa, Botswana and Namibia, as well as in smaller scattered populations in Zimbabwe, Zambia and Angola. The fact that the San generally lived in small groups in remote locations added to their isolation, and contributed towards their vulnerability to exploitation by others.

Generally impoverished, marginalized and cut off from the modern world, they received minimal support from their respective governments. Almost no communication took place between the leaders of these far-flung communities, with the result that their ability to share information and empower their peoples remained structurally constrained.

---

[2] The following are the most common major San languages currently spoken in the region. Botswana hosts Nharo, Gwi, G/anna and Khwe; Namibia hosts Ju/huasi, Hei//om, Kung, !Xun and Khwe; South Africa hosts the !Khomani, the !Xun and the Khwe; Zimbabwe hosts the Tyua.

The fate suffered by the San peoples in Africa is similar to that of many indigenous peoples in other parts of the world. Expansion and conquest, firstly by assertive local pastoralist and agriculturalist communities, followed later and with similar devastation by colonial powers, all but obliterated their former existence. The San history over the centuries has been one of dispossession, enslavement, cultural extinction and recorded patterns of officially sanctioned genocide (Penn 2013).

For many reasons, including their lifestyle until recent times as hunter-gatherer peoples, and their unique genetic properties as descendants of possibly the earliest members of the human race, the San have found themselves in high demand as research populations.

Modern San leaders faced with increasing societal challenges had no means of communicating their problems with other leaders, of learning about their human rights, or of discussing ways in which they might legitimately challenge the unwanted interventions of researchers and other outsiders such as media practitioners.

In addition, the San world view is generally one of seeking harmony, and avoiding all forms of conflict. Several scholars of conflict resolution have based their principles of good practice on ancient San systems, in which the prevention of disputes and the reconciliation of interests are deeply ingrained (Ury 1995).

The outside world regarded the San as a classic example of a "vulnerable population", lacking the means to organize a collective expression of their common interests and concerns (Chennells 2009). Prior to the year 2000, virtually all research was externally conceived, and was perceived by the San as being disruptive and on occasion harmful to the research populations (Chennells and Steenkamp 2018).

Internet searches of the words San, Khoisan[3] and Bushmen throw up thousands of papers, books and research theses, supporting the assertion that they are among the most researched peoples in the world. Until they formed their own representative organizations, they did not have a unified voice and thus remained powerless to resist unwanted attention from outsiders.

Institution Building and Supportive NGOs

The most important step towards the San Code of Research Ethics was local institution-building, an initiative that made all further successes possible.

---

[3] While the term "Khoisan" is frequently used in general discourse as a collective name for two distinct groupings in southern Africa, namely the Khoi, or KhoiKhoi, and the San, this umbrella term is not relevant to a discussion of the San peoples. The Khoi or KhoiKhoi, formerly known in South Africa as Hottentots, are regarded as pastoral, and of more recent origin (Barnard 1992).

WIMSA: The Catalyst Institution

WIMSA (the Working Group of Indigenous Minorities in Southern Africa) has arguably been the most important of a number of San support organizations operating in southern Africa over the past 25 years. Reverend Mario Mahongo, one of the San leaders whose work on the San Code of Research Ethics was crucial, noted of a 1996 workshop: "For the first time we were meeting San leaders from the whole region, and we realised that this new organisation WIMSA could really help our people" (Chennells and Schroeder 2019).

Table 7.1 lists the main non-governmental organizations (NGOs) that have provided services to the San in South Africa, Botswana and Namibia.

Supported through seed funding from Swedish and Dutch charities, German development worker Axel Thoma and San leaders such as Kipi George and Augustino Victorino promoted the formation of a cross-border, regional organization to protect the rights of all San peoples in southern Africa. The topics that emerged as clear priorities among San communities were:

- Access to land
- Benefit sharing for traditional knowledge
- Protection of heritage and culture[4]

WIMSA functioned effectively as a regional organization from its inception in 1996 until approximately 2016. The successes of this important San organization in raising awareness and promoting advocacy among the San cannot be overstated.

Table 7.1 San support organizations

Start year	Organization name	Organization region
1981	Nyae Nyae Development Foundation	Tsumkwe, Namibia
1988	Kuru Development Trust	Ghanzi, Botswana
1991	First People of the Kalahari	Ghanzi, Botswana
1992	First Regional San Conference	Windhoek, Namibia
1995	Final Regional San Conference (pre-WIMSA)	D'Kar, Botswana
1996	WIMSA	Windhoek, Namibia
1996	South African San Institute (SASI)	Kimberley, South Africa
1999	!Khwa ttu San Culture and Education Centre	Darling, South Africa.
2001`	South African San Council	Upington, South Africa
2006	Namibia San Council	Windhoek, Namibia
2007	Khwedom Council	Gaborone, Botswana.

---

[4]A San organisation that is highly active and successful in protecting the San cultural heritage is !Khwa ttu, which was not directly involved in the development of the San Code of Research Ethics and is therefore not included here with its own section. Details about !Khwa ttu can be found in Chennells and Schroeder (2019).

Importantly, in 1998 WIMSA drafted the San's first Media and Research Contract, which was aimed at managing external incursions into San culture which up to that time had occurred with no San control at all. San leaders throughout the region were trained in the implementation of the contract, and it was used to deal with researchers, filmmakers, writers and others who entered San territory wanting to gather information.

South African San Institute and South African San Council

The San Code of Research Ethics takes a step further than the WIMSA Media and Research Contract. It outlines exactly what the San require from researchers. The WIMSA contract, by contrast, is more akin to an ethics approval form, which requires researchers to provide information about their studies before they enter San communities. One major difference is therefore that the San code requires collaboration from the start – that is, from the inception of the research – rather than approving fully conceived studies as through the WIMSA form.

The two NGOs most important in developing the San Code of Research Ethics were the South African San Institute (SASI) and the South African San Council (SASC).

SASI was formed in 1996 and initially took the form of a dedicated San service NGO. SASI's original mission was to assist the !Khomani San with their restitution land claim in the Kalahari. This was completed successfully in 1999, but SASI continued to be active. SASI also supported the !Xun and Khwe San communities, who were relocated to South Africa from Namibia after the end of the "bush wars" in 1990, and settled in a temporary army camp near Kimberley, where SASI is based. The communities' first needs were for assistance in relation to housing and other social problems arising from their exceedingly disrupted and war-torn history, having been caught in the crossfire between the apartheid government of South Africa and guerrilla fighters in Angola and Namibia. SASI was the partner in the TRUST project which represented the San peoples, and which assisted with the development of the San Code of Ethics. They hosted all relevant workshops and the launch of the code in Cape Town (see below).

The SASC had existed informally since 1996, representing the interests of three South African San communities on the WIMSA board (!Khomani, Khwe, !Xun). It was legally constituted in 2001 so that it could negotiate officially on behalf of the San, and proceeded over the years to become a major success story in San institution building. The SASC negotiated a famous benefit-sharing agreement with South Africa's Council for Scientific and Industrial Research (CSIR) in relation to the San's traditional knowledge rights to the Hoodia plant.

The global UN Convention on Biological Diversity (CBD) of 1992 was the first instrument to provide for the principle that commercial users of plants with active ingredients based upon traditional knowledge needed to negotiate benefit-sharing agreements with the holders of the traditional knowledge, in order to ensure fairness. With this development, the San rediscovered the value of their culture and heritage in the form of their traditional knowledge of a wide range of medicinal and other useful indigenous plants. In 2003 the first benefit-sharing agreement was concluded with the CSIR.

The Hoodia benefit-sharing case achieved seminal status in the CBD world (Wynberg et al. 2009) and can be seen as the first major step taken by the San towards achieving fairness in research. The strong demand for benefit sharing in research is also the reason why fairness is an important, separate value in the San Code of Research Ethics.

The importance of the collaboration of SASI (support NGO) and the SASC (representing San issues directly through San leaders) has been emphasized by the SASC's director, Leana Snyders:

> Our relationship with SASI has helped increase our capacity to understand the law, and also to represent our people. With the legal knowledge gained from negotiating benefit sharing agreements resulting from our traditional knowledge, the San have become acknowledged leaders in this field. (Chennells and Schroeder 2019)

Leaders of Integrity

It is perhaps a truism that collective progress is impossible without leaders of vision and integrity. When the San began their process of institutional development in 1996, they were fortunate to have a group of pioneering leaders who drove and supported the vision to end the isolation of the past and to enter the organizational modern world. The San were blessed during this period with strong leaders, some of whom are still active, who had the wisdom to support change and the ability to engender consensus among sometimes differing opinions while retaining the confidence and trust of their people.

One can be forgiven, however, for singling out the following leaders, who died prematurely while dedicated to the process of empowering their people: Kipi George (Khwe), /Xau Moses (Ju//Huansi), Augustino Victorino (!Xun), Robert Derenge (Khwe), Dawid Kruiper (!Khomani), Andries Steenkamp (!Khomani), and Mario Mahongo (!Xun).

These leaders rose above their peers for many reasons, including the following, which are drawn from the many eulogies delivered upon their passing: they were strong and able to take difficult decisions, without losing an element of softness and humanity; each was regarded as honest and dedicated to his people, rather than to his immediate family and clan; they were respected both by their own communities and by outsiders for their intelligence, integrity and wisdom. These factors alone made them unique, and, like Nelson Mandela, they are constantly invoked as icons of leadership.

The two San leaders who contributed most to the San Code of Research Ethics were Andries Steenkamp and Mario Mahongo. Two messages to researchers made by them have meanwhile achieved iconic status:

> I don't want researchers to see us as museums who cannot speak for themselves and who don't expect something in return. As humans we need support. (TRUST Project Global Research Ethics 2018b)
> – Reverend Mario Mahongo (1952–2018)

> Your house must have a door so that nobody needs to come in through the window. You must come in via the door, that is to say via the San Council. (TRUST Project Global Research Ethics 2018a)
> – Andries Steenkamp (1960–2016)

The last statement has even made it into the San Code of Research Ethics (see below), which notes:

> Andries Steenkamp, the respected San leader who contributed to this Code of Ethics until he passed away in 2016, asked researchers to come through the door, not the window. The door stands for the San processes. When researchers respect the door, the San can have research that is positive for us.

The leaders who have succeeded Andries and Mario are focusing on many unresolved questions, in particular:

1. Why do some researchers still come into the San community through the window, like thieves? For instance, are they not aware of the community structures? Do they not trust the structures? Is there intent to avoid community approval?
2. How can awareness of the San Code of Research Ethics be spread throughout the far flung San communities in South Africa and potentially into Botswana and Namibia? How in a practical sense and how in a financial sense?
3. How can the on-line approval and code adherence system that the SASC wishes to install be designed and funded, both for development and for maintenance?
4. Could the San effort be captured in a model fit to assist other communities that do not have a 25-year history of institution building around their rights?
5. As the San community wishes to assist others in developing their own codes, how can such efforts be funded?

Legal Support

Many of the important steps undertaken along the path of community empowerment require legal support or intervention. The formulation of constitutions, leases and basic legal documents underpinning salaried appointments, and the drafting of basic agreements with government, funders and other external actors all require the services of a lawyer to protect the San's interests.

WIMSA and SASI have, from the outset, retained the services of an in-house lawyer. This ensures that they receive basic institutional legal support, as well as strategic legal support, in their various advocacy programmes. Apart from basic institutional legal support, the most visible advocacy successes of the San have all relied upon close collaboration with a legal adviser.

San policy interventions at the United Nations, land claims and successful San claims for intellectual property rights related to their traditional knowledge (on Hoodia, buchu, Sceletium, rooibos etc), which raised the international profile of the San as indigenous peoples, all required committed legal support. This was made available mostly via SASI.

The prohibitive cost of standard commercial lawyers is a well-known deterrent to obtaining legal advice and assistance. In addition, utilizing lawyers who are not familiar with the ethos and needs of the community can lead to expensive mistakes and misunderstandings. Lawyers who are willing to represent the community legally on a pro bono or noncommercial basis can therefore give a vulnerable community a significant advantage.

Dr Roger Chennells, SASI's lawyer, also provided a legal editing service for the San Code of Research Ethics.

Supportive Research Collaborations

Formulating ethics codes is a time-consuming business that requires funding, in particular to support workshops where San traditional leaders and San community members can discuss their concerns and ways forward. Sceptics may point out that the same individuals always attend such workshops largely out of appreciation for the food provided, and leave without any tangible or lasting benefits.

By contrast, there is much anecdotal evidence of San colleagues who reported, after attending workshops, that their thinking, and indeed sometimes their lives, had forever been altered by an insight gained at the workshop. The San development programmes conducted by WIMSA, SASI and the SASC held capacity-building workshops on a range of topics. Of particular relevance to the San Code of Research Ethics were the workshops funded by two successive EU projects, ProGReSS[5] and TRUST.[6]

The ProGReSS project, under the leadership of Professor Doris Schroeder, ran from 2013 to 2016 with SASI as a partner with its own budget. The project funded two workshops to revise the WIMSA Research and Media Contract, among other things. By the conclusion of ProGReSS, it became clear that a new San ethics code might realize San interests more effectively in the future.

The EU-funded TRUST project, also led by Professor Schroeder, catalysed a global collaborative effort to improve adherence to high ethical standards in research around the world. Its main product is the Global Code of Conduct for Research in Resource-Poor Settings, the main topic of this book. However, a second high-profile output is the San Code of Research Ethics.

Both projects enabled productive workshops to be held, at which the San's rights were further debated, and where the outcomes were not only used by the San in

[5] http://www.progressproject.eu/

[6] http://trust-project.eu/

practical cases, but also published and disseminated. The TRUST project united the efforts of many years to tackle the challenge of how unwanted research could be controlled. Without the collaborative support of international research partners, it is doubtful that the San Code of Research Ethics would have emerged.

Drafting the San Code of Research Ethics

Building on various earlier efforts, the San Code of Research Ethics was drafted over the course of three workshops and much intervening work during the year prior to its launch in March 2017.

In March 2016, SASI organized a preparatory workshop at which San representatives voiced their concerns and reported their past involvement in national and international research studies. Examples of good and bad research case studies were identified, in order to guide a revision of the San Media and Research Contract and the drafting of a San Code of Research Ethics. The aim was to help the South African San manage their involvement in research and heritage studies. Delegates included SASC members plus leaders from the !Xun, the Khwe and the !Khomani, together with selected invited experts from the fields of genetics, sociology, ethnology, research ethics and law. During this workshop the participants received background information on research in the different fields, delivered by the experts attending the workshop.

Based on this input, initial ideas to improve research engagement were developed. The following ideas were voiced:

- A single central body needs to be created with clear external and internal authority, and the capacity to manage research and media issues.
- A code of ethics needs to be established, whereby researchers are able to understand the "dos and don'ts" of engaging with the San.
- Training needs to take place, both of the leaders or local coordinators of research and among the communities and individuals who are required to participate.
- Research and media contracts need to be drawn up in such a way that research is not discouraged, but is managed for the benefit of the community. Research which is not felt to be useful should be refused.
- Noncommercial research or engagement should be managed with basic contracts. More in-depth research should be managed with more complex contracts as appropriate.
- There should be consequences and penalties for failure to comply with the terms of such contracts.
- Funds should be raised in order to establish a research monitoring and compliance body with the SASC.

In May 2016, SASI organized a full workshop with 22 San representatives and a further eight external contributors, again from the fields of genetics, sociology, ethnology, research ethics and law. On this occasion, work was undertaken to ensure that the San could protect themselves from exploitation in research through the redrafting of their original research contract, and by the development of what was to become the San Code of Research Ethics.[7]

During this important workshop, the San developed a range of general principles that applied to their own community. These principles were as follows and were used for a first draft of the San Code of Research Ethics:

- The San require respect to the environment, to San leaders and individuals, and to cultural values.
- Honesty, integrity and honour are important between all partners.
- Cultural and spiritual values must be fully honoured and respected in all research and media projects.
- The right formal process should be followed to protect communities in research.
- Informed consent is central to all research.
- Genetics samples should only be used for the purpose stated in the research contract.
- Researchers should not enter a community without being guided and led by members of the community itself.
- Both researcher and community should benefit from the interaction.

In November 2016, SASI organized a third workshop with the same delegates and some of the earlier external contributors to finalize the content of the San Code of Research Ethics. The overall goal was to achieve fair research partnerships. The following threats and weaknesses were discussed.

- Vulnerable and far-flung populations and serious poverty
- Undue influence by researchers, due to poverty
- "Free riders" who do not support San community concerns when taking part in research for cash
- Exploitation possibilities due to illiteracy
- Lack of knowledge of research, what it means and what its risks are
- Lack of knowledge about the San leadership's approach to research
- Lack of assistance from the government
- Low self-esteem in engaging with outside individuals and agencies
- Earlier theft of traditional knowledge leading to mistrust of researchers
- Lack of system to combat the problems
- Lack of institutional and financial support to the leadership who aim to improve the situation

With these challenges in mind, the initial draft of the San Code of Research Ethics was revised and refined. In addition, each element of the new draft code was grouped into one of the four TRUST ethical values of fairness, respect, care, and honesty. These values had been agreed on previously by the TRUST group, with San input. The four core values were to be supported by a fifth value, which the San

---

[7]A short video presentation about the workshop is available at https://www.youtube.com/watch?v=HOdw3mv7JSo.

delegates deemed essential, namely proper process. In small groups the key points of each value were written out in greater detail.

A highly important decision was that examples of past exploitation would form part of the code itself.

The results of the third workshop were then given to colleagues who undertook further work. In December 2016, Roger Chennells worked on the code from a legal perspective, and in January 2017, Doris Schroeder worked on its ethical dimensions.

The subsequent draft, which had been edited from both a legal and an ethical perspective, was then presented to the San leadership for adoption. Further minor changes were made, until the code was unanimously adopted and declared ready to be launched by the San leadership.

The San Code of Research Ethics

Respect

We require respect, not only for individuals but also for the community.

We require respect for our culture, which also includes our history. We have certain sensitivities that are not known by others. Respect is shown when we can input into all research endeavours at all stages so that we can explain these sensitivities.

Respect for our culture includes respect for our relationship with the environment.

Respect for individuals requires the protection of our privacy at all times.

Respect requires that our contribution to research is acknowledged at all times.

Respect requires that promises made by researchers need to be met.

Respectful researchers engage with us in advance of carrying out research. There should be no assumption that San will automatically approve of any research projects that are brought to us.

We have encountered lack of respect in many instances in the past. In Genomics research, our leaders were avoided, and respect was not shown to them. Researchers took photographs of individuals in their homes, of breastfeeding mothers, or of underage children, whilst ignoring our social customs and norms. Bribes or other advantages were offered. Failure by researchers to meet their promises to provide feedback is an example of disrespect which is encountered frequently.

Honesty

We require honesty from all those who come to us with research proposals.

We require an open and clear exchange between the researchers and our leaders. The language must be clear, not academic. Complex issues must be carefully and correctly described, not simply assuming the San cannot understand. There must be a totally honest sharing of information.

Open exchange should not patronise the San. Open exchanges implies that an assessment was made of possible harms or problems for the San resulting from the research and that these possible harms are honestly communicated.

Prior informed consent can only be based on honesty in the communications, which needs to be carefully documented. Honesty also means absolute transparency in all aspects of the engagement, including the funding situation, the purpose of the research, and any changes that might occur during the process.

Honesty requires an open and continuous mode of communication between the San and researchers.

We have encountered lack of honesty in many instances in the past. Researchers have deviated from the stated purpose of research, failed to honour a promise to show the San the research prior to publication, and published a biased paper based upon leading questions given to young San trainees. This lack of honesty caused much damage among the public, and harmed the trust between the collaborating organisation and the San. Another common lack of honesty is exaggerated claims of the researcher's lack of resources, and thus the researchers' inability to provide any benefits at all.

Justice and Fairness

We require justice and fairness in research.

It is important that the San be meaningfully involved in the proposed studies, which includes learning about the benefits that the participants and the community might expect. These might be largely non-monetary but include co-research opportunities, sharing of skills and research capacity, and roles for translators and research assistants, to give some examples.

Any possible benefits should be discussed with the San, in order to ensure that these benefits do actually return to the community.

As part of justice and fairness the San will try to enforce compliance with any breach of the Code, including through the use of dispute resolution mechanisms.

In extreme cases the listing and publication of unethical researchers in a "black book" might be considered.

An institution whose researchers fail to comply with the Code can be refused collaboration in future research. Hence, there will be "consequences" for researchers who fail to comply with the Code.

We have encountered lack of justice and fairness in many instances in the past. These include theft of San traditional knowledge by researchers. At the same time, many companies in South Africa and globally are benefitting from our traditional knowledge in sales of indigenous plant varieties without benefit sharing agreements, proving the need for further compliance measures to ensure fairness.

Care

Research should be aligned to local needs and improve the lives of San. This means that the research process must be carried out with care for all involved, especially the San community.

The caring part of research must extend to the families of those involved, as well as to the social and physical environment.

Excellence in research is also required, in order for it to be positive and caring for the San. Research that is not up to a high standard might result in bad interactions, which will be lacking in care for the community.

Caring research needs to accept the San people as they are, and take note of the cultural and social requirements of this Code of Ethics.

We have encountered lack of care in many instances in the past. For instance, we were spoken down to, or confused with complicated scientific language, or treated as ignorant. Failing to ensure that something is left behind that improves the lives of the San also represents lack of care.

Process

Researchers need to follow the processes that are set out in our research protocols carefully, in order for this Code of Ethics to work.

The San research protocol that the San Council will manage is an important process that we have decided on, which will set out specific requirements through every step of the research process.

This process starts with a research idea that is collectively designed, through to approval of the project, and subsequent publications.

The San commit to engaging fairly with researchers and manage effectively all stages of the research process, as their resources allow. They also commit to respecting the various local San structures (e.g. Communal Property Association, CPA leaders) in their communications between San leaders and San communities.

Andries Steenkamp, the respected San leader who contributed to this Code of Ethics until he passed away in 2016, asked researchers to come through the door, not the window. The door stands for the San processes. When researchers respect the door, the San can have research that is positive for us.

Conclusion

Key to the achievements of the San in South Africa have been: dedicated San leaders of integrity, supportive NGOs, legal support, and long-term relationships with key individuals who also assisted with fundraising (see Fig. 7.1).

Fig. 7.1 Success factors

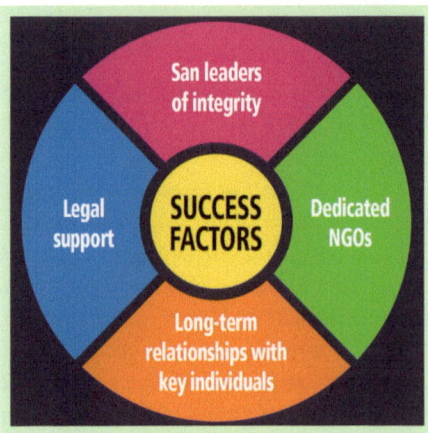

The San leadership have developed an approach to outsiders, for instance researchers, that is open to forging authentic human relationships. Every research project meeting or benefit-sharing negotiation was regarded as an opportunity to meet a certain person who might prove himself or herself to be mutually open to a relationship of trust.

In particular Andries Steenkamp of the !Khomani San and Mario Mahongo of the !Xun San, both former chairpersons of the SASC, formed such relationships of trust. Not only was the famous San humour seldom far from the surface, but they exuded an air of confidence and open curiosity, quick to understand and appreciate the persons across the table, and slow to take personal offence. Their personal integrity shone through, and the trust that they generated in others translated into untold benefits for the San.

This approach ensured that the San Council is highly respected in South Africa. In addition, relationships of trust developed with international researchers, generating funding for research and policy projects. One of the many results of the openness of San leaders to collaboration with the world is the San Code of Research Ethics, which, it is hoped, will put all future relationships with outsiders onto an equitable basis. As Leana Snyders put it:

> The San Code of Research Ethics is the voice of a community that have been exploited for so many years. This code manages to bridge the gap between the research community and the San Community through dialogue. By taking ownership of the code, the San Community will ensure that this document will remain relevant for generations to come. (Chennells and Schroeder 2019)

References

Barnard A (1992) Hunters and herders of southern Africa: a comparative ethnography of the Khoisan peoples. Cambridge University Press, New York and Cambridge

Callaway E (2017) South Africa's San people issue ethics code to scientists. Nature 543:475–476. https://www.nature.com/news/south-africa-s-san-people-issue-ethics-code-to-scientists-1.21684

Chennells R (2009) Vulnerability and indigenous communities: are the San of South Africa a vulnerable people? Cambridge Quarterly of Healthcare Ethics 8(2):147–154

Chennells R, Steenkamp A (2018) International genomics research involving the San people. In: Schroeder D, Cook J, Hirsch F, Fenet S, Muthuswamy V (eds) Ethics dumping: case studies from North–South research collaborations. Springer, Berlin, p 15–22

Chennells R, Schroeder D (2019) The San Code of Research Ethics: its origins and history, a report for TRUST. http://www.globalcodeofconduct.org/affiliated-codes/

Hitchcock RK, Ikeya K, Biesele M, Lee RB (2006) Introduction. In: Hitchcock RK, Ikeya K, Biesele M, Lee RB (eds) Updating the San: image and reality of an African people in the 21st century. Senri Technological Studies 70. National Museum of Ethnology, Osaka, p 4

Knight A, Underhill PA, Mortensen HM, Zhivotovsky, LA, Lin AA, Henn BM, Louis D, Ruhlen M, Mountain JL (2003). African Y chromosome and mtDNA divergence provides insight into the history of click languages. Current Biology 13(6):464–473

Penn N (2013) The British and the 'Bushmen': the massacre of the Cape San, 1795 to 1828. Journal of Genocide Research 15(2):183–200 https://doi.org/10.1080/14623528.2013.793081

TRUST Project Global Research Ethics. (2018a) Andries Steenkamp and Petrus Vaalbooi interviews – TRUST Project. https://www.youtube.com/watch?v=A4_Mvdwl_Gc

TRUST Project Global Research Ethics. (2018b) Reverend Mario Mahongo – TRUST Project. https://www.youtube.com/watch?v=jMhCUNw9eAo

Soodyall H (2006) A prehistory of Africa. Jonathan Ball Publishers, Jeppetown, South Africa

Ury W (1995) Conflict resolution amongst the Bushmen: lessons in dispute systems design. Harvard Negotiation Journal 11(4): 379–389

Wynberg R. Schroeder D, Chennells R (2009) Indigenous peoples, consent and benefit sharing: lessons from the San-Hoodia Case. Berlin, Springer

Chapter 8
Good Practice to Counter Ethics Dumping

Abstract An ethics code is not enough to avoid ethics dumping. Ethics codes can inspire, guide and raise awareness of ethical issues, but they cannot, on their own, guarantee ethical outcomes; this requires a multifaceted approach. For research in resource-poor settings, engagement is crucial. Such engagement has been built into the Global Code of Conduct for Research in Resource-Poor Settings as a requirement, but how can it be put into practice? An approach for ethical community engagement is presented in this chapter, which also includes suggestions for an accessible complaints mechanism. At the institutional level, we tackle the question of concluding fair research contracts when access to legal advice is limited. Throughout, at a broader level, we show how the four values of fairness, respect, care and honesty can be used to help guide decision-making and the practical application of the code.

Keywords Community engagement · Complaints procedure · Research contracts · Values compass

Introduction

The Global Code of Conduct for Research in Resource-Poor Settings (GCC) is not enough to prevent ethics dumping[1]. While codes are necessary, they are not sufficient in themselves to ensure good governance (Webley and Werner 2008). For codes to be effective, researchers must know how to use them appropriately (Giorgini et al. 2015), and codes can be totally ineffective when badly implemented (Bowman 2000). The use of codes, especially new ones, invariably raises challenges of interpretation and implementation.

The way that the GCC has been developed helps to minimize potential challenges. For instance, implementation problems are lessened when the needs, values

---

[1] The export of unethical research from a high-income setting to a resource-poor setting with weaker compliance structures or legal governance mechanisms.

© The Author(s), under exclusive license to Springer Nature Switzerland AG 2019   89
D. Schroeder et al., *Equitable Research Partnerships*, SpringerBriefs in Research and Innovation Governance, https://doi.org/10.1007/978-3-030-15745-6_8

and interests of all stakeholders,[2] particularly those who are likely to be subject to the codes (Lawton 2004), are taken into account during development. This helps ensure that the code is aligned with real needs and has practical value. Hence the bottom-up approach that was taken during development, which has facilitated an "insider" perspective, should help to increase effectiveness and counter the view that ethics codes are no more than a bureaucratic tool imposed from above. Additionally, the GCC does not replicate existing codes, nor does it seek to replace them. Rather, the GCC can be viewed as *complementary* to other codes, and this helps to avoid the confusion that can arise when codes seem contradictory.

However, even the most conscientiously developed codes are open to differences in interpretation, and researchers need an ethical foundation for making decisions about application in particular situations (Eriksson et al. 2008). Furthermore, codes must form part of a wider framework that also includes mechanisms for compliance, accountability and addressing legal concerns.

This chapter seeks to address these issues with practical guidance for implementation. We show how the four values of *fairness*, *respect*, *care* and *honesty* can serve as an ethical foundation for decision-making. Specifically, we highlight the importance of ethical techniques for engagement with local communities and a complaints procedure which is accessible to highly vulnerable populations, and finally we summarize a resource built in parallel to the code: a fair research contracting tool.

The Values as an Ethical Foundation

The four values constitute the foundation for ethical research collaborations and can be applied in virtually any situation to guide decision-making. When researchers keep the values at the heart of their activities, they can recognize and respond to ethical challenges more effectively. This requires reflexivity[3] on the part of the researchers such that they consciously and regularly "stand back" from their activities to ask whether their activities are aligned with the values. At any stage researchers must ask themselves: *Am I behaving with fairness, respect, care and honesty?* We call this practical application of the values the "values compass" (see Fig. 8.1).

The compass can be used continually as a tool for ethical reflection, but is particularly helpful at key stages of the research process when important decisions are made.

---

[2] As noted earlier, "stakeholders" is an increasingly contested term, as it may imply that all parties hold an equal stake. Some prefer the term "actors", yet this brings its own complexities. While acknowledging the debate, we use the well-established term "stakeholders" throughout.

[3] "Reflexivity" can be thought of as a researcher's ongoing critical reflection upon his or her own biases and assumptions and how these impact upon their relationship to the research, the course of the research and knowledge production.

Fig. 8.1 The values
compass

In the following sections we show how the practical application of the values can help guide two important activities in collaborative research: community engagement and the development of an accessible complaints procedure.

Ethical Engagement with Communities

The term "community" is contentious and contextual, and can be difficult to define (Day 2006). For the purposes of this chapter, we use an early definition from the World Health Organization which describes a community as:

> A specific group of people, often living in a defined geographical area, who share a common culture, values and norms, are arranged in a social structure according to relationships which the community has developed over a period of time. Members of a community gain their personal and social identity by sharing common beliefs, values and norms which have been developed by the community in the past and may be modified in the future (WHO 1998: 5)

As we can infer from this definition, there are many different types of communities and also communities within communities. For example, indigenous communities, having a historical continuity with preinvasion and precolonial societies that developed on their territories, may consider themselves distinct from other sectors of the societies that now prevail on those territories, or parts of them. They generally form nondominant sectors of society and can be intent on preserving, developing and transmitting to future generations their ancestral territories and their ethnic identity, as the basis of their continued existence as peoples, in accordance with their own cultural patterns, social institutions and legal systems (Martínez Cobo

2014). They often have particular relationships with advocacy groups who work to protect or represent their interests.[4]

The concept of communities within communities also includes groups of people who are vulnerable because of a range of physical (disabilities, for example) or cultural (religion, for example) characteristics. For instance, sex workers, injecting drug users and men who have sex with men are often marginalized within their own broader communities.[5] People from such groups are frequently sought for international research and yet the community at large or the community leaders are often unable to provide the input needed to ensure ethical management of research projects. Communities and their leaders may be unaware of the specific circumstances of these people and their lives, and they may even be openly hostile. We therefore need mechanisms for ensuring that the voice of marginalized and vulnerable populations is heard, and that their interests in research are represented.

In the 1990s, community engagement assumed prominence as the new guiding light of public health efforts; research and health-improvement programmes that involved communities had better results than programmes led by government alone (NIH 2011). At the same time, the limitations of existing guidelines for the protection of communities in genetic research was becoming increasingly apparent (Weijer et al. 1999). The benefits of community engagement in all types of research are now widely acknowledged, and numerous publications describe many potential benefits such as:

- increasing community understanding and acceptance of the studies
- enhancing researchers' ability to understand and address community priorities
- improving logistics and the running of studies
- strengthening the quality of the information collected
- ensuring culturally sensitive communications and research approaches
- enhancing opportunities for capacity building (Hebert et al. 2009; Cook 2008; Bassler et al. 2008; Dunn 2011).

Community engagement is an ethical imperative (a "must") for researchers operating globally. Research participants, their local communities and research partners in international locations should be equal stakeholders in the pursuit of research-related gains (Anderson et al. 2012). Ahmed and Palermo (2010) provide a salient definition of community engagement in research as

> a process of inclusive participation that supports mutual respect of values, strategies, and actions for authentic partnership of people affiliated with or self-identified by geographic proximity, special interest, or similar situations to address issues affecting the well-being of the community of focus.

To be effective in international research, community engagement requires the development of partnerships with "local" stakeholders (for example, national,

---

[4] Advocacy groups (also known as pressure groups, lobby groups, campaign groups, interest groups or special interest groups) use various forms of advocacy in order to influence public opinion and/ or policy.

[5] Here "broader community" can refer to a village, town, ethnic group etc.

regional or advocacy groups), involving them in assessing local challenges and research priorities, determining the value of research, planning, conducting and overseeing research, and integrating the results with local needs where relevant (Jones and Wells 2007). Moreover, it requires members of the research team to become part of the community, and members of the community to become part of the research team to create bespoke working environments before, during and after the research.

Many models have been proposed for effective community engagement in research,[6] and many written guides already exist. Rather than add an invention of our own to the numerous existing models, we show here how reference to the four values of fairness, respect, care and honesty can highlight the primary ethical considerations for organizations or researchers engaging with communities over the course of a research project. After all, as Dunn (2011: 5) points out, "Engagement is not a benchmark for ethics. Ethics does not stop when community engagement takes place. Engagement itself has ethical implications."

Our guidance for community engagement is intended to be useful; we show how application of the values compass at key stages of the research process can invoke particular questions for contemplation. There may be other relevant questions, depending upon the circumstances, but these questions are a useful starting point.

The key stages we consider are:

1. Setting the research aims and/or developing the research question
2. Designing the study
3. Implementing the study
4. The results phase
5. Evaluating the study

Setting the Research Aims and/or Developing the Research Question

During the initial phase, when researchers are formulating their research aims or a research question, the values compass can be applied to help ensure that ethical considerations are attended to during their community engagement. Table 8.1 enumerates checklist questions for contemplation at the very outset.

One of the first tasks is to establish community preferences and protocols for engagement, even before discussions about the research are started. This may require a local spokesperson or other trusted intermediary to be identified.

---

[6] Examples are community-based participatory research, empowerment evaluation, community action research and participatory rapid appraisal.

Table 8.1 Questions for reflection when setting research aims

Fairness	Honesty
How are the community being meaningfully involved in discussions about the aims of the research including why it is needed and who will benefit?	Have all background details been shared and discussed with the community, including the funding situation and the intentions of the researchers?
	What procedures will be used for two-way, open communication?
	What procedures are in place to ensure, without being patronizing, that research issues are understood?
	What promises are being made to the community, and can they definitely be fulfilled?
Respect	**Care**
How are community preferences for engagement strategies being discussed and acted upon?	How are local needs and the potential for capacity building being taken into account in developing the aims of the study?
Are the relevant community spokespersons or representatives being consulted?	
Is permission from community elders/leaders or representatives needed for this consultation?	Is due attention being paid to the impact of the study and the study team upon the participants, their families, the local community and the environment?
How are the research team familiarizing themselves with local culture – including organizational structures, history, traditions, relationship with the environment and sensitivities?	

Designing the Study

The community, as well as local researchers, need to be included in the research design process, both effectively and transparently. Table 8.2 sets out some of the many factors to consider.

It is imperative that researchers consider the practical implications for the persons, communities and environments involved, as well as the scientific integrity of the research design.

Implementing the Study

Ethical research is conducted *with* communities rather than *about* communities. To ensure that this is how it actually happens, effective engagement is vital throughout the implementation of the project. Table 8.3 suggests questions to consider during the implementation phase.

Table 8.2 Questions for reflection when designing a study

Fairness	Honesty
How are the community involved in the planning and design of the study?	How is full transparency in all aspects of the engagement and planning being ensured?
Are the potential benefits and harms for the participants and the community being discussed fully?	Are procedures for open, two-way communication in place?
Have the most relevant types of benefits for the participants and communities been discussed and agreed?	Have all details that might impact upon individuals or the community been disclosed?
In health research, has post-study access to successfully tested treatments or interventions been agreed?	Have requirements for an accessible and user-friendly complaints mechanism been discussed and agreed?
Where relevant, have means for recognizing and protecting traditional knowledge been agreed?	What promises are being made to the local community in the design of the study and are they likely to be fulfilled?
Respect	**Care**
Are the research team complying with local/community ethics codes?	How are local needs being taken into account in the design of the study?
How is community knowledge being respected and integrated into the design?	Is due attention being paid to the impact of the study and the study team upon the participants, their families, the local community and the environment?
Are the relevant members of the community, as identified by the community itself, involved in the design?	What measures have been taken to ensure understanding (such as translators and the use of clear, non-technical language)
How is community culture and tradition being respected in the design of the study?	Have the resource implications of this design for the local community been identified?
Have the relevant persons in the community given permission/approval for the study design?	What measures are in place to ensure that the research is high quality and worthwhile so that the efforts of the community are not wasted?

Where possible, members of the local community should be actively involved in undertaking the research. This may be in simple, practical operational or administrative capacities, but where appropriately qualified or experienced candidates are available, and/or where necessary training can be provided, this involvement should also include more complex tasks, with support from experienced researchers.

Table 8.3 Questions for reflection during implementation

Fairness	Honesty
How are the local community engaged in the ongoing implementation of the research?	How are lines of communication functioning? Is there clear and transparent, two-way communication between the research team and the local community?
Are local researchers and other members of the community taking active roles in the implementation?	
Have measures for ensuring ethical compliance been discussed with the community and put in place?	How are the community being informed about developments or any changes that occur during the research process?
	How is the complaints system functioning? Does it need to be amended in any way?
Respect	**Care**
Are researchers taking steps to ensure all activities are respectful of local culture and traditions?	Have the researchers taken the time and necessary steps to ensure that the implications of the study have been fully understood by participants and the community?
Has both individual and community consent, assent or approval (where appropriate) been granted?	Are researchers paying due attention to the impact of the study and the study team upon the participants, their families, the local community and the environment?
What measures are in place to respect rights to privacy, anonymity and confidentiality?	Is the community being properly resourced for participation?
Are the participants and community fully aware of their right to withhold personal/ sensitive information and to refuse engagement/participation?	

The Results Phase

During this phase, results are analysed and disseminated through publications as well as being fed back to the community. Table 8.4 formulates some helpful questions to ask during the results phase.

Findings can be enriched when members of the community have been consulted and engaged during the analysis process and the interpretation of results. For some studies, sharing results with the research participants or the community can elucidate aspects that were previously obscure to the researchers (for example, an understanding of why or how something happens).

Evaluating the Study

Though the publication of research results and feedback to the community represent, in a sense, the end of the research cycle, the process of further research involving the same or other communities can be greatly helped by an evaluation of the

Table 8.4 Questions for reflection during analysis of results

Fairness	Honesty
How are members of the local community involved in analysis and interpretation of the results?	Have promises that were made about access to the results been fulfilled?
What measures are in place to ensure access to findings that might be beneficial to the community?	Have all findings been disclosed in an honest manner?
Are appropriate steps being taken to recognize and protect traditional knowledge contributions?	
Respect	**Care**
Have the community been given an opportunity to review the results and implications of the study prior to publication?	What measures are in place to ensure that the findings and implications of the study are accessible to and fully understood by participants and the community?
Have the community's knowledge and contribution been fully acknowledged in the results?	
Have community culture and tradition been taken into consideration in the interpretation of the results?	
Have rights to privacy, anonymity and confidentiality in reporting been respected?	

study and, in particular in the context of this report, of the community involvement elements. Table 8.5 lists important questions for the evaluation phase.

When researchers apply the values of fairness, respect, care and honesty over the course of a research project, this creates a relationship of trust with the community. Our main advice for ethical community engagement is to build long-term, mutually beneficial relationships based on the four values, applied *before, during* and *after* research studies.

In addition, for a relationship of trust to develop between researchers and local communities, it is important to have a well-functioning complaints procedure

Developing an Accessible Complaints Procedure

The routine use of accessible complaints procedures in research forms part of the overarching strategy for reducing ethics dumping, because such procedures can help to ensure that experience and practice correspond with expectations. An effective complaints procedure can give voice to those who participate in research, offering a channel for raising concerns that might otherwise remain unheard, both during and after a study. Complaints procedures can contribute to the safeguarding of

Table 8.5 Questions for reflection during the evaluation of a study

Fairness	Honesty
Have the agreed benefits for participation been realized?	Have all promises to the community been fulfilled?
In health research, is the agreed post-study access to successfully tested treatments or interventions being made available?	How have complaints been managed? Are there lessons to be learned and shared?
	Have implications that might impact upon individuals or the community, including potential harms and benefits, been disclosed?
How have the community been involved in the evaluation of the research *findings*?	
How have the community been involved in the evaluation of the research *process*?	
Do the community believe that they have benefited from the research?	
Respect	**Care**
Are there mechanisms in place to feedback news about broader impacts of the research?	Do the community believe that researchers paid due attention to the impact of the study and the study team upon the participants, their families, the local community and the environment?
Has the contribution of members of the local community been fully credited?	
Have the community's knowledge and its value to the research been fully credited?	Was the resulting project of high quality and worthwhile so that the efforts of the community were not wasted?
Do the community believe that local culture and tradition have been respected?	

participants[7] so that it endures beyond the ethical approval process; they offer a mechanism for correcting mistakes and for protecting people, animals and the environment from abuse and mistreatment. Significantly, complaints mechanisms offer a means of revealing lapses and failures in ethical conduct, thereby providing opportunities for enhancing ethical compliance in research (Fig. 8.2).

Researchers, research organizations and research ethics committees (RECs) can go to great lengths to ensure that research protocols are scientifically rigorous and that research is conducted in accordance with the relevant ethical principles. However, even when the greatest care is taken, unexpected events can occur and participation can lead to emotional and/or physical harm. While most RECs will specify the need for an identified contact person in case of queries or complaints, this commonly takes the form of basic contact details on a participant information sheet, often in the form of an email address. Where further information is given, it

---

[7] "Participation" is referred to here in its broadest sense to include experimental research animals and environments, as well as human research participants, local communities and researchers.

Fig. 8.2 The functions of an effective complaints procedure

frequently stipulates that all complaints must be made in writing. The requirement to complain in writing via email might preclude complaints from the most vulnerable research participants.

For collaborative research undertaken in resource-poor settings, especially low and middle-income countries (LMICs), the accessibility of a complaints procedure may be affected by many factors that are unfamiliar to researchers from a high-income country (HIC). A concerted effort is therefore required to understand local needs and preferences so that a complaints mechanism can be implemented that is both user-friendly and fit for purpose.

Factors Affecting Accessibility

It is known from studies in the field of dispute resolution that people often feel reluctant to make complaints and that this can be related to a variety of complex factors. In 2009 the Health Professions Council in the UK published a comprehensive scoping review of existing mechanisms for complaints about health professionals (HPC 2009). In this report, the HPC describes a range of factors that can act as barriers to making a specific complaint. As there are no equivalent publications about complaints procedures in LMICs, we summarize here the factors that are relevant to research in LMICs.

Readiness to complain in any environment can be influenced by gender, ethnicity, age, education, income, accessibility of information and the perceived "seriousness" of the problem (Pleasence et al. 2006). Specifically, ethnic minority communities are less likely to use systems that they perceive as being culturally insensitive and are more fearful of the consequences of taking action when they feel those systems have failed them.

Difficulties with access to information are highlighted as a barrier to making a complaint (Henwood et al. 2003), especially where there is "information illiteracy"; some people possess the relevant skills and confidence to seek out information, but many do not. In situations where levels of education and literacy are not high, this is likely to be exacerbated.

The relationship between the person who brings the complaint and the bureaucracies to which they must direct their complaint can be a factor (Cowan and Halliday 2003). This relationship can either encourage or discourage a potential complainant's trust in complaints mechanisms. The power imbalance between parties in such relationships can be substantial. For example, when working with impoverished communities, HIC researchers should be aware that participation in a clinical study may provide a participant's only access to health care or other much-needed benefits. Fear of retribution is often cited as a barrier to making a complaint, particularly in circumstances where the complainant has an ongoing relationship with the complainee (HPC 2009). In situations where there is a power imbalance, people may not have the confidence to complain; they may be reluctant to seem ungrateful, not wish to be seen as a complainer, or fear loss. Research has shown that some people even reconstruct negative experiences in a positive light in order to maintain relationships (Edwards et al. 2004).

In addition to the above, participatory engagement activities in the TRUST project (Chapter 6) have revealed the following factors that could also act as barriers to research participants making complaints about research activities in LMICs:

- *Fear of damage or stigmatization from loss of confidentiality or anonymity*. In Kenya, for example, where sex work is illegal, sex workers may be reluctant to make any formal complaints.
- *Cultural norms that preclude complaining*. In some cultures, it is not acceptable to make complaints, especially to or about visitors and/or those in authority. Complaining may be perceived as disrespectful, ungrateful or inappropriate.
- *Illiteracy of research participants and communication (language) difficulties*, leading to a lack of understanding of reasonable rights relating to informed consent and to reasonable expectations of the research.

- *Inability to access the means by which to file a complaint*: for example, if only an email address is provided as a contact and one has no access to computers or internet connections.

The Scope of a Complaints Procedure

A comprehensive complaints procedure can have a broad scope; it can be used to complain about any activities that are associated with a research study. These may include, for example:

- any perceived deviation from the information provided
- any deviation from agreed processes
- treatment by members of the research team that is considered inappropriate
- problems with the organization of the study (for example, the competence of the researchers and their ability to perform duties)
- the (mis)handling of personal or sensitive information
- concerns about any unethical behaviour or practices by the research team

The scope of a complaints procedure will also depend upon the intended users. Many complaints procedures are intended for use purely by participants in a research study. However, in collaborative ventures in LMICs, there may be a wide range of potential users, because HIC-LMIC collaborative research is especially prone to ethics dumping, with the potential for damage to entire communities.

Table 8.6 gives examples of the potential range of users of complaints procedures for different types of research studies.

While a complaints procedure can have broad scope, it is vital that there be clarity about its purpose and who can use it, as well as about what can and cannot be dealt with through this mechanism. A lack of common understanding of any procedure's purpose can be a source of great dissatisfaction and cause wider distrust in the process.

Table 8.6 Potential users of complaints procedures in different types of research

Social science	Clinical trials	Animal experimentation	Agricultural research
Research participants	Research participants	Local community	Local farmers
Local community	Local community	Local researchers	Broader local community
Local researchers	Local researchers	Local animal handlers	Local researchers
Local research organizations	Local research organizations	Local animal research centres	

A Values-Based Approach to Developing a Complaints Procedure

A complaints procedure that works perfectly well in one location and for one purpose cannot simply be transposed to a different situation without due consideration of its applicability. Local relevance and accessibility are vital keys in the design of an effective complaints procedure. Rather than a formally laid-down set of "rules" for complaints procedures, a strategic values-based approach needs to be implemented to deal with different levels and types of complaints, so that individuals and communities feel respected, cared for, fairly treated, fully informed and empowered. The four values can provide a framework for the development of an appropriate procedure, as shown in Table 8.7.

Any complaints procedure for a research study involving LMIC populations, especially vulnerable groups or communities, must first consider the circumstances, situation and culture of such communities and the individuals to be recruited to the study. A critical step in this process is engagement with the community that will be

Table 8.7 Values-based considerations for the development of a complaints procedure

Fairness	Respect
Responses to complaints should be timely	The procedure for complaints should be respectful of local needs and preferences
All complaints should be taken seriously and investigated fully	
Records of complaints and responses should be maintained to enable reporting and monitoring of complaints	Appropriate levels of confidentiality and privacy should be maintained throughout the procedure (including for all documentation, investigations, discussions and hearings)
The nature and types of redress should be acceptable to the local community	Researchers and/or appropriate staff should be fully equipped and trained for implementation of the complaints procedure.
The lodging of honest complaints should be encouraged, and even facilitated, in order to overcome power imbalances.	
Honesty	**Care**
The purpose and limitations of the complaints procedure should be clearly communicated to all involved in the research	The local community should be involved at an early stage in the development of the complaints procedure
The process for making a complaint should be clearly communicated to all involved in the research	Advice should be taken from the local community about the accessibility and viability of the complaints procedure. This may mean offering a range of methods for information sharing and complaint acceptance – verbal, written, and through trusted spokespersons and community groups etc.
This process should be as simple and straightforward as possible	

involved with or affected by the research so that they can help guide the development of appropriate procedures.

Additionally, strategies[8] will need to be developed for dealing with different types of complaints. It is important to try to avoid complex and overly burdensome strategies which all too easily become legalistic and formalized. In practice this can mean that nothing is set up at all, or that what is established becomes little more than an ineffective bureaucratic exercise. While more formal approaches and structures may work in "Western" settings, these are unlikely to be effective in the kinds of vulnerable communities where care is needed to safeguard and empower; they may even have the opposite effect, and discourage any engagement at all on complaints issues.

Equally, the challenges in establishing an effective strategy should not act as an excuse for researchers to adopt an oversimplified model (such as a contact name on the information sheet) that is of little or no benefit to anyone. For each unique situation, researchers should work with communities to cocreate effective strategies that take into account the circumstances, situation and culture of that community and the individuals to be recruited to the study.

While it is not possible for us to specify a single "model" complaints procedure, we have shown how the values can provide the basis of any complaints procedure. With these values embedded in the thinking of the research community, they can then seek to work with whatever procedures and structures are available, adapting, improving and tailoring them for application in the real world. The individuals and groups involved should feel respected, cared for, fully informed, treated fairly and empowered.

Most protective mechanisms, including complaints procedures, are strengthened when supported by legal systems, but participants, communities, researchers and institutions in LMICs often have no or very limited access to legal advice or protection. The next section introduces an online toolkit that will be helpful in such situations.

A Fair Research Contracting Tool

The need for fair research contracts is best illustrated by the situation in international collaborative health research. Research undertaken in LMICs can lead to significant benefits flowing into HICs. In 2009, Glickman et al. undertook a systematic review to examine what had led to a "dramatic shift in the location of clinical trials" and concluded that important factors were:

- shortened timelines for clinical testing due to a larger pool of research participants

---

[8] These might include internal resolution through study-specific schemes; internal resolution through research ethics committees; litigation through the courts; or alternative dispute resolution mechanisms such as mediation, adjudication and arbitration.

- lower regulatory barriers for research in LMICs
- international harmonization of intellectual property rights protection

To take full advantage of the benefits of conducting medical research in LMICs, research institutions in HICs have invested substantially in building legal and contracting expertise for the benefit of their own institutions and stakeholders. Such expertise may not be as easily available in LMIC institutions. As a result, the benefits of research collaborations remain heavily skewed towards the beneficiaries based in HICs (Sack et al. 2009).

In 2011 the Council on Health Research for Development (COHRED) committed itself to launching its Fair Research Contracting (FRC) initiative to support LMIC partners when negotiating equitable research partnerships. FRC aimed to identify best practices for the research contracting process that would be useful in the following three scenarios:

- where there is no lawyer
- where there may be lay personnel who could be trained
- where there is a lawyer or legal expertise

A basic framework was subsequently developed by COHRED and partners to assist LMIC collaborators in making contractual demands on HIC collaborators without requiring large legal teams of their own. This focused on the fair distribution of post-research benefits, intellectual property rights, data and data ownerships, specimen ownership and usage, technology transfer and institutional capacity building as key outcomes of the FRC process. Between 2015 and 2018, and as part of the TRUST project, the existing FRC framework was enhanced and expanded to provide an online toolkit relevant for all types of research.

The FRC online toolkit[9] now provides information, tips and case studies in six key areas:

- Negotiation strategies: for understanding the various aspects of negotiations, whether a research partner is at a basic starting point or an advanced level in the development of contract negotiations
- Research contracting: for a basic understanding of contracts and contracting so that a research partner can better manage responsibilities, opportunities and risks that impact the research partnership
- Research data: providing the essential principles concerning rights and responsibilities, including accountability and access to data in collaborative research
- Intellectual property: providing an introduction to some of the key general principles that require consideration before participation in collaborative research agreements
- Research costing: providing research partners with a basic understanding of cost considerations when developing a full cost research budget proposal

---

[9] The entire online toolkit is available at http://frcweb.cohred.org/

- Technology transfer and capacity: concerning the flow of knowledge, experience and materials from one partner to another, and the ability of people and organizations to manage their affairs and reach objectives successfully.

The development of this resource means that vulnerable groups, such as communities or researchers without legal support, have access to resources that can help develop a good understanding of research contracting for equitable research partnerships and avoid exploitation in research.

Conclusion

According to Eriksson et al. (2008), a serious flaw in most new ethics guidelines is that they are produced with the pretension that there are no other guidelines in existence, and it would be much better if they just stated what they *added* to existing guidelines. Such is the case with the GCC, which focuses solely on factors that are specific to collaborative research ventures in resource-poor (primarily LMIC) settings. The GCC is succinct and written in plain language; it is meant to be equally accessible to researchers in HICs and to their intended partners in LMICs. In these respects, the GCC is very straightforward, but its simplicity will inevitably generate questions about how it should be implemented.

For example, article 13 of the GCC states that a clear procedure for feedback, complaints or allegations of misconduct must be offered that gives genuine and appropriate access to all research participants and local partners to express any concerns they may have with the research process. Aside from the injunction that the procedure must be agreed with local partners at the outset of the research, there is no guidance on what this procedure should look like. This "thin approach" was used for a reason: no complaints mechanism will fit all situations. Hence, the emphasis is on the *process*, namely to agree with local partners on an approach. Codes are not enough in themselves to ensure ethical conduct; they need buy-in from all those involved, and such buy-in needs to be generated through effective engagement mechanisms.

Researchers should therefore see community engagement as the gateway to effective implementation of the GCC. For example, when considering the local relevance of the proposed research (article 1), who better to ask than members of the local community? When wondering how best to seek informed consent, who better to ask than members of the local community? Consultation with the community offers the most direct route to addressing questions about implementation and to realizing the essence of the GCC: a global collaborative effort to eradicate ethics dumping.

References

Ahmed SM, Palermo AGS (2010) Community engagement in research: frameworks for education and peer review. American Journal of Public Health 100(8):1380–1387. https://doi.org/10.2105/AJPH.2009.178137

Anderson EE, Solomon S, Heitman E, DuBois JM, Fisher CB, Kost RG, Ross LF (2012) Research ethics education for community-engaged research: a review and research agenda. Journal of Empirical Research on Human Research Ethics 7(2):322–319

Bassler A, Brasier K, Fogel N, Taverno R (2008) Developing effective citizen engagement: a how-to guide for community leaders. Center for Rural Pennsylvania, Harrisburg PA. http://www.rural.palegislature.us/effective_citizen_engagement.pdf

Bowman JS (2000) Towards a professional ethos: from regulatory to reflective codes. International Review of Administrative Sciences 66:673–687

Cook WK (2008) Integrating research and action: a systematic review of community-based participatory research to address health disparities in environmental and occupational health in the United States. Journal of Epidemiology and Community Health 62(8):668–676. https://doi.org/10.1136/jech.2007.067645

Cowan D, Halliday S (2003) The appeal of internal review: law, administrative justice and the (non-) emergence of disputes. Hart, Oxford

Day G (2006) Community and everyday life. Routledge, London

Dunn A (2011) Community engagement: under the microscope. Wellcome Trust, London

Edwards C, Staniszweska S, Crichton N (2004) Investigation of the ways in which patients' reports of their satisfaction with healthcare are constructed. Sociology of Health and Illness 26(2):159

Eriksson S, Höglund AT, Helgesson G (2008) Do ethical guidelines give guidance? A critical examination of eight ethics regulations. Cambridge Quarterly of Healthcare Ethics 17(1):15–29

Giorgini V, Mecca JT, Gibson C, Medeiros K, Mumford MD, Connelly S, Devenport LD (2015) Researcher perceptions of ethical guidelines and codes of conduct. Accountability in Research 22(3):123–138

Glickman SW, McHutchison JG, Peterson ED, Cairns CB, Harrington RA, Califf RM, Schulman KA (2009) Ethical and scientific implications of the globalization of clinical research. New England Journal of Medicine 360:816–823. https://doi.org/10.1056/NEJMsb0803929

Hebert JR, Brandt HM, Armstead CA, Adams SA, Steck SE (2009) Interdisciplinary, translational, and community-based participatory research: finding a common language to improve cancer research. Cancer Epidemiology, Biomarkers & Prevention 18(4):1213–1217. https://doi.org/10.1158/1055-9965.EPI-08-1166

Henwood F, Wyatt S, Hart A, Smith, J (2003) Ignorance is bliss sometimes: constraints on the emergence of the "informed patient" in the changing landscapes of health information. Sociology of Health and Illness 25(6):589–607

HPC (2009) Scoping report on existing research on complaints mechanisms. Health Professions Council. https://www.hcpc-uk.org/resources/reports/2009/scoping-report-on-existing-research-on-complaints-mechanisms/

Jones L, Wells K (2007) Strategies for academic and clinician engagement in community-participatory partnered research. Journal of the American Medical Association 297(4):407–410

Lawton A (2004) Developing and implementing codes of ethics. Viešoji politika ir administravimas 7:94–101

Martínez Cobo M (2014) Study on the problem of discrimination against indigenous populations. United Nations Department of Economic and Social Affairs. https://www.un.org/development/desa/indigenouspeoples/publications/2014/09/martinez-cobo-study/

NIH (2011) Principles of community engagement. Washington, DC: CTSA Community Engagement Key Function Committee Task Force on the Principles of Community Engagement, National Institutes of Health. https://www.atsdr.cdc.gov/communityengagement/pdf/PCE_Report_508_FINAL.pdf

Pleasence P, Buck A, Balmer N, O'Grady A, Genn H, Smith M (2006) Causes of action: civil law and social justice. The Stationery Office, Norwich

Sack DA, Brooks V, Behan M, Cravioto A, Kennedy A, IJsselmuiden C, Sewankambo N (2009) Improving international research contracting. WHO Bulletin 87:487–488

Webley S, Werner A (2008) Corporate codes of ethics: necessary but not sufficient. Business Ethics: A European Review 17(4):405–415

Weijer C, Goldsand G, Emanuel EJ (1999) Protecting communities in research: current guidelines and limits of extrapolation. Nature Genetics 23(3):275

WHO (1998) Health promotion glossary. World Health Organization, Geneva. http://www.who.int/healthpromotion/about/HPR%20Glossary%201998.pdf, page 5.

Chapter 9
Towards Equitable Research Partnership

Abstract The world's largest collection of professional ethics codes already holds more than 2,500 codes. What can the Global Code of Conduct for Research in Resource-Poor Settings (GCC) add? This brief chapter gives co-authors and supporters of the GCC the opportunity to show why a code with the single-minded aim of eradicating ethics dumping is needed.

Keywords Global ethics · Research ethics · International co-operation · Ethics dumping · Low- and middle-income countries

The world's largest collection of professional ethics codes holds more than 2,500 codes (IIT nd). Is another ethics code really needed? The evidence gathered on the 21st-century export of unethical research practices from high-income to lower-income settings says it is (Schroeder et al. 2018). Such ethics dumping still occurs despite a proliferation of ethics codes. This could be for either of two main possible reasons. First, an ethics code designed to guard against ethics dumping is not yet available. Second, ethics codes are not suitable for guarding against ethics dumping. We would agree with the first, but not the second.

Instead of summarizing the book, this chapter concludes by giving the floor to co-authors and supporters of the Global Code of Conduct for Research in Resource-Poor Settings (GCC) and the San Code of Research Ethics. Their statements provide further evidence that both codes are truly needed and that they will make a difference in the campaign against ethics dumping.[1]

---

[1] Quotes below without cited references are taken from personal written communications addressed to Doris Schroeder in early 2019.

We found that many research stakeholders expressed a need for a simpler, more intuitive and fairer framework to guide research practice. – *Professor Carel IJsselmuiden (South African), executive director of the COHRED Group, co-author of the GCC*

The San Code of Research Ethics is the voice of a community that have been exploited for so many years. The San community will ensure that this document will remain relevant for generations to come. – *Leana Snyders (South African), director of the South African San Institute, co-author of both codes*

We want to know whether our global project is going well – ethically. The GCC provides a short and accessible checklist to prompt reflection on our stroke care research in hospitals in India. – *Professor Dame Caroline Watkins (British), leader of the Improvise Project, first individual project adopter of the GGC*

The GCC is a good guide for Indian ethics committees and the Health Ministry's Screening Committee to review Indo-EU collaborative studies, including those by Indian researchers with EU researchers. – *Dr Nandini Kumar (Indian), vice president of the Forum for Ethics Review Committees in India, co-author of the GCC*

The four values of the GCC at last answer the perpetual question that has nagged people who care about acting ethically: how do you make individuals act ethically in a world where there are too many codes, and ethics dumping still happens? Inspire them! The four global values of fairness, respect, care and honesty inspire individuals in any context to act ethically. – *Professor Pamela Andanda (Kenyan), professor of law at the University of the Witwatersrand, South Africa, co-author of the GCC*

The inclusion in the GCC of elements relating to agricultural and environmental ethics is long overdue. It is time that fairness, respect, care and honesty were considered more systematically in agriculture. – *Associate Professor Rachel Wynberg (South African), South African Bio-economy Research Chair, co-author of the GCC*

I have had a very close look at the Global Code of Conduct that you have proposed, and I really find it impressive (Burtscher 2018). – *Wolfgang Burtscher (Austrian), deputy director-general of Research and Innovation for the European Commission*

Unlike other codes, the GCC has built into it community engagement and meaningful involvement of research participants as part of a checklist for designing good studies. – *Dr Joshua Kimani (Kenyan), clinical research director at Partners for Health and Development in Africa, University of Manitoba field office in Kenya, co-author of the GCC*

Many of my company's clients are not ethicists, but they want to undertake their research ethically. The GCC avoids ethics jargon and is concise and clear. – *Elena Tavlaki (Greek), director of Signosis Sprl, co-author of the GCC*

Good practices know no regional or political boundaries: research that is unethical in Europe is unethical in Africa. That's why the GCC is needed. – *Dr Michael Makanga (Ugandan), executive director of the European & Developing Countries Clinical Trials Partnership, co-author of the GCC*

The world is unfair. We are talking about R&D in an unfair world ... The Code of Conduct [GCC] is exquisitely clear that it is unethical to do research in one place for the sake of another. (TRUST 2018) – *Professor Jeffrey Sachs (American), speaking at the GCC launch in the European Parliament*

The new four-values system around fairness, respect, care and honesty is highly appreciated in Asia. People find it intuitive – in fact, most audiences loved it. – *Dr Vasantha Muthuswamy (Indian), president of the Forum for Ethics Review Committees in India, co-author of the GCC*

Zhai Xiaomei [Chinese], the executive director of the Centre for Bioethics at the Chinese Academy of Medical Sciences, in Beijing, who is also deputy director of the health ministry's ethics committee, welcomes what TRUST[2] has done. (Economist 2018)

To deliver our mission to end world hunger, we need to undertake research. Applying the GCC will assist us greatly. No previous code was designed so clearly for work with highly vulnerable populations in resource-poor settings. – *Myriam Ait Aissa (French), head of Research and Analysis at Action contre la Faim, co-author of the GCC*

Fairness, respect, care and honesty: four simple words with clear meaning to help researchers enter the house through the door and no longer through the window.[3] – *Dr François Hirsch (French), former head of the Inserm (French National Institute of Health and Medical Research) Office for Ethics, co-author of the GCC*

Ron Iphofen [British], an adviser on research ethics to the European Commission, believes the code will have a profound impact on how funding proposals to the EU are designed and reviewed. "I could envisage reviewers [of EU-funded research proposals] now looking suspiciously at any application for funds that entailed research by wealthy nations on the less wealthy that did not mention the code," he says. (Nordling 2018)

The emphasis in the GCC on fairness, respect, care and honesty resonates with our work at UNESCO. – *Dr Dafna Feinholz (Mexican), UNESCO's chief of Bioethics and Ethics of Science and Technology, co-author of the GCC*

---

[2] The EU-funded consortium that developed the GCC.

[3] This refers to Andries Steenkamp's iconic request to researchers, namely to enter San communities through the metaphorical "front door" – that is, the San Council – and not, like thieves, through the window.

TRUST was a game changer.[4]

Ethics dumping is a real threat to the quality of science and the GCC is now a mandatory reference document for EU framework program funding to guard against it. – *Dorian Karatzas (Greek), head of Ethics and Research Integrity, European Commission*

Best science for the most neglected, also means best ethical standards. That's why the GCC aims high: to protect the most neglected. – *Dr François Bompart (French), director of Paediatric HIV/Hepatitis C Programmes at the Drugs for Neglected Diseases initiative (DNDi), former vice president, Access to Medicines at Sanofi, co-author of the GCC*

We get given consent forms and documents, often in a hurry. We sign because we need the money and then end up with regret. It feels like a form of abuse. They want something from us and they know how to get it. Because of our socio-economic conditions, we will always be vulnerable to those from the North. A code of ethics is needed that protects indigenous people.[5] – *Andries Steenkamp (1960–2016) (South African), former chair of the South African San Council, co-author of both codes*

I don't want researchers to see us as museums who cannot speak for themselves and who don't expect something in return. As humans, we need support.[6] – *Reverend Mario Mahongo (1952–2018) (Angolan), co-author of both codes*

We want to be treated by researchers with fairness, respect, care and honesty. Is that too much to ask?[7] – *Joyce Adhiambo Odhiambo (Kenyan), health activist and former sex worker, co-author of the GCC*

Indeed, is that too much to ask?

References

Burtscher W (2018) TRUST Global Code of Conduct to be a reference document applied by all research projects applying for H2020 funding. TRUST eNewsletter Issue 5. http://www.global-codeofconduct.org/wp-content/uploads/2018/12/TRUSTNewsletter_2018_Issue5.pdf
Economist (2018) Recent events highlight an unpleasant scientific practice: ethics dumping. The Economist, 31 January. https://www.economist.com/science-and-technology/2019/02/02/recent-events-highlight-an-unpleasant-scientific-practice-ethics-dumping
IIT (nd) The ethics codes collection. Center for the Study of Ethics in the Professions, Illinois Institute of Technology, Chicago. http://ethicscodescollection.org/

---

[4] Opening words during European Commission ethics staff training on the GCC, 3 Dec 2019, Covent Garden Building, Brussels.

[5] Recorded message from Nairobi TRUST meeting, 23 May 2016.

[6] Recorded video message from Kimberley TRUST meeting, Feb 2017.

[7] European Parliament, TRUST event, 29 June 2018.

Nordling L (2018) Europe's biggest research fund cracks down on "ethics dumping". Nature 559:17–18. https://www.nature.com/articles/d41586-018-05616-w

Schroeder D, Cook J, Hirsch F, Fenet S, Muthuswamy V (eds) (2018) Ethics dumping: case studies from North-South research collaborations. Springer Briefs in Research and Innovation Governance, Berlin

TRUST (2018) Major TRUST event successfully held at European Parliament. TRUST eNewsletter Issue 5. http://www.globalcodeofconduct.org/wp-content/uploads/2018/12/TRUSTNewsletter_2018_Issue5.pdf

Appendix

Global Code of Conduct for Research in Resource-Poor Settings (GCC)

Authors and Sources of Inspiration

Lead Author: **Doris Schroeder** University of Central Lancashire, UK
    Authors in alphabetical order:

- **Joyce Adhiambo** Partners for Health and Development, Kenya
- **Chiara Altare** Action contre la Faim, France
- **Fatima Alvarez-Castillo** University of the Philippines, Philippines
- **Pamela Andanda** University of the Witwatersrand, South Africa
- **François Bompart** Drugs for Neglected Diseases Initiative, Switzerland
- **Francesca I. Cavallaro** UNESCO, France
- **Kate Chatfield** University of Central Lancashire, UK
- **Roger Chennells** South African San Institute, South Africa
- **David Coles** University of Central Lancashire, UK
- **Julie Cook** University of Central Lancashire, UK
- **Julia Dammann** South African San Institute, South Africa
- **Amy Azra Dean** (media) University of Central Lancashire, UK
- **Dafna Feinholz** UNESCO, France
- **Solveig Fenet** Institut national de la santé et de la recherche médicale, France
- **François Hirsch** Institut national de la santé et de la recherche médicale, France
- **Carel IJsselmuiden** Council on Health Research for Development, Switzerland
- **Sandhya Kamat** Forum for Ethics Review Committees in India, India
- **Rosemary Kasiba** Partners for Health and Development, Kenya
- **John Kiai** Partners for Health and Development, Kenya
- **Joshua Kimani** Partners for Health and Development, Kenya
- **Mihalis Kritikos** Scientific Foresight Unit, European Parliament, Belgium

- **Olga Kubar** Saint-Petersburg Pasteur Institute, Russia
- **Nandini K Kumar** Forum for Ethics Review Committees in India, India
- **Miltos Ladikas** University of Central Lancashire, UK
- **Klaus M Leisinger** Foundation Global Values Alliance, Switzerland
- **Collin Louw** South African San Council, South Africa
- **Gwenaelle Luc** Action contre la Faim, France
- **Mario Mahongo** South African San Council, South Africa
- **Michael Makanga** European and Developing Countries Clinical Trials Partnership, the Netherlands
- **David Morton** University of Birmingham, UK
- **Ngaya Munuo** University of Cape Town, South Africa
- **Vasantha Muthuswami** Forum for Ethics Review Committees in India, India
- **Peter Mwaura** Partners for Health and Development, Kenya
- **Dieynaba N'diaye** Action contre la Faim, France
- **Jaci van Niekerk** University of Cape Town, South Africa
- **Catherine Njoki** Partners for Health and Development, Kenya
- **Holger Postulart** Council on Health Research for Development, Switzerland
- **Johannes Rath** International biosafety and security adviser (EU, UN)
- **Zeka Shiwarra** South African San Council, South Africa
- **Karin M Schmitt** Foundation Global Values Alliance, Switzerland
- **Miriam Shuchman**, University of Toronto, Canada
- **Michelle Singh** European and Developing Countries Clinical Trials Partnership, South Africa
- **Giorgio Sirugo** University of Pennsylvania, USA
- **Leana Snyders** South African San Council, South Africa
- **Andries Steenkamp** South African San Council, South Africa
- **Hennie Swart** South African San Institute, South Africa
- **Elena Tavlaki** Signosis, Belgium
- **Urmila Thatte** Forum for Ethics Review Committees in India, India
- **Jacintha Toohey** Council on Health Research for Development Association, Switzerland
- **Anthony Tukai** Partners for Health and Development, Kenya
- **Josephine Waithera** Partners for Health and Development, Kenya
- **Jane Wathuta** University of the Witwatersrand, South Africa
- **Paul Woodgate** Wellcome Trust, UK
- **Rachel Wynberg** University of Cape Town, South Africa
- **Yandong Zhao** Chinese Academy of Science and Technology for Development, China

We gratefully acknowledge valuable input on the GCC from the following research funders:

- Wellcome Trust
- Global Forum on Bioethics in Research
- Calouste Gulbenkian Foundation
- Medical Research Council UKRI

- World Health Organization TDR
- European Commission.

We also received valuable input on the GCC from the following industry partners:

- Sanofi
- Roche
- Novartis
- GlaxoSmithKline
- Boehringer Ingelheim
- European Federation of Pharmaceutical Industries and Associations

Existing guidelines played an important role in the formulation of the GCC. While the code does not quote any articles directly, it draws substantial inspiration from the ethics guidelines listed below.

- Canadian Coalition for Global Health Research: Principles for Global Health Research
- CIOMS and WHO: International Ethical Guidelines for Health-related Research Involving Humans
- Convention on Biological Diversity: Nagoya Protocol
- Council on Health Research for Development: Research Fairness Initiative
- European Code of Conduct for Research Integrity
- GlaxoSmithKline: Clinical Trials in the Developing World
- Indian Council of Medical Research: National Ethical Guidelines for Biomedical and Health Research Involving Human Participants
- International Society of Ethnobiology: ISE Code of Ethics
- Research Councils UK: RCUK Common Principles on Data Policy
- Roche: Animal Research – Roche Principles of Care and Use
- Sanofi: Corporate Social Responsibility Factsheet on Biodiversity and Biopiracy
- South African Medical Research Council. Use of Animals in Research and Training
- South African San Institute, South African San Council and TRUST project: San Code of Research Ethics
- Swiss Academy of Sciences, Commission for Research Partnerships with Developing Countries: 11 Principles & 7 Questions. KFPE's Guide for Transboundary Research Partnerships
- 3rd World Conference on Research Integrity: Montreal Statement on Research Integrity in Cross-boundary Research Collaborations
- UNESCO: Universal Declaration on Bioethics and Human Rights 2005
- World Medical Association: Declaration of Helsinki – Ethical Principles for Medical Research Involving Human Subjects

The GCC was developed using a mission statement (Fig. 1).

Fig. 1 Mission statement of the GCC authors

Index